PREFACE

ARC WELDING is intended for beginning students in welding. It deals specifically with welding principles and practices in three major areas—shielded metal arc, gas tungsten arc (Tig), and gas metal arc (Mig).

ARC WELDING consists of fifteen units. Each unit includes basic information which welding students should know. The *Self Quiz* following each unit is designed to help students find out how well they understand welding principles.

Most units also contain a variety of welding assignments. These assignments are intended to give students an opportunity to develop welding skills.

By reading ARC WELDING and by completing the assignments at the end of each unit students will not necessarily become experienced welders. However, they can gain enough knowledge and skill to become employable in industry and to be able to progress more rapidly in work demanding greater skill and responsibility.

J.W. Giachino

CONTENTS

Unit   Page

1 Introduction to Welding 1
2 Reading Weld Symbols 9
3 Basic Welding Metallurgy 19
4 Shielded Metal-Arc Electrodes 31
5 Shielded Metal-Arc Welding Equipment 39
6 Principles of Shielded Metal-Arc Welding 50
7 Shielded Metal Arc—Flat Position Welding 59
8 Shielded Metal Arc—Horizontal Welding 72
9 Shielded Metal Arc—Vertical Welding 78
10 Shielded Metal Arc—Overhead Welding 84
11 Gas Tungsten-Arc Welding 90
12 Gas Metal-Arc Welding 112
13 Pipe Welding 136
14 Arc Cutting 149
15 Certification of Welders 153
Answers to Self Quiz Items 160

TK4660 .G494

Giachino, Joseph William, 1906-

Arc welding

arc welding

Joseph W. Giachino

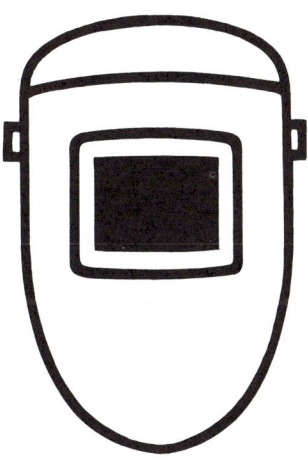

AMERICAN TECHNICAL PUBLISHERS, INC.
HOMEWOOD, ILLINOIS 60430

COPYRIGHT © 1977

BY AMERICAN TECHNICAL PUBLISHERS, INC.

Library of Congress Catalog Number: 76-051153
ISBN: 0-8269-3083-2

No portion of this publication may be reproduced by *any* process such as photocopying, recording, storage in a retrieval system or transmitted by any means without permission of the publisher

123456789-77-9

PRINTED IN THE UNITED STATES OF AMERICA

UNIT 1
Introduction to Welding

Welding is essential to the expansion and productivity of our industries. Welding has become one of the principal means of fabricating and repairing metal products. It is almost impossible to name an industry, large or small, that does not employ some type of welding. Industry has found that welding is an efficient, dependable, and economical means of joining metal in practically all metal fabricating operations and in most construction.

Occupational Opportunities in Welding

The widespread use of welding in American industry provides a constant source of employment for both skilled and semi-skilled operators. According to U.S. Department of Labor there are approximately 555,000 persons employed as welders. Three-fifths of these work in industries that manufacture durable goods, such as transportation equipment, machinery, and household products. Most of the others work for construction firms and repair shops.

Employment outlook. Employment of welders is expected to increase because of the development of newer and better welding processes. This is particularly true in ship building, tank and boiler fabrication, rail, automotive, and aircraft manufacture, building construction, piping, and many other metalworking industries. Although there is no uniform wage rate for welders, they are considered to be in one of the higher classifications of job wages.

Training. Learning the essential skills needed to fulfill the many welding job requirements varies from a few months of on-the-job training to several years of formal training. Most employers prefer applicants who have a high school education or vocational school training in welding. See Fig. 1-1. Courses in mathematics, mechanical drawing, blueprint reading and general metals are very helpful.

Job classification. A beginner usually starts on simple production jobs and gradually works up to higher levels of skill as his experience and ability improve. Before being assigned to work where the quality and strength of the weld are critical, a welder will generally have to pass a certification test given by the employer, government agency or some other inspection authority.

Welders are usually classified as skilled and semi-skilled. Skilled welders are those who have the ability to plan, lay out work from drawings or written specifications, and weld all types of joints in various positions, such as flat, vertical, horizontal and overhead. They also have a wide

2 Arc Welding

Fig. 1-1. To attain the required welding skills, a person usually has to complete a formal course of instruction under a competent instructor.

range of technical knowledge involving properties of metals, effects of heat on welded structures, control of expansion and contraction forces, reading welding symbols, and recognizing welding defects. The skilled welder may be proficient in several welding areas encompassing gas and arc welding processes. As a rule the skilled welder is always certified for the particular welding job he is required to perform.

The semi-skilled welders usually do repetitive work, that is, production work which generally does not involve critical safety and strength requirements. They primarily weld surfaces in only one position and may or may not have to be certified.

The following are some of the principal job titles of welders:

Welding engineer
Arc welder—shielded metal-arc
—gas-shielded arc
Welding cutter
Submerged-arc welder
Welder—resistance, spot, automatic
Pipe welder
Boilermaker welder
Structural welder
Maintenance welder
Welding layout and set-up man
Welding inspector
Welding tester
Welding foreman-supervisor

Skilled welders may, by promotion, become inspectors, foremen, or supervisors. Actually there are unlimited opportunities for those who become thoroughly acquainted with the techniques, materials, designs, and new applications of welding processes.

Types of Welding Processes

The principal welding processes in use today are classified as shielded metal arc, gas tungsten arc, gas metal arc, oxy-acetylene and resistance.

Shielded metal arc is a welding process in which fusion is achieved by an arc generated between a consumable, coated electrode and the work.

Gas tungsten-arc is a welding process in which an arc is formed between a nonconsumable tungsten electrode and the work under a protective shield of gas or gas mixture.

Gas metal-arc is a welding process in which fusion is achieved by means of an arc formed between a continuous, consumable wire electrode and the work under a protective shield of gas or gas mixture.

Oxy-acetylene is a welding process in which welding is done with a gas flame, produced by burning a mixture of oxygen and acetylene, and a filler rod.

Resistance welding consists of a group of welding processes (spot, seam, pulsation, projection) where welding is done with the heat produced by the resistance of the work to the passing of an electric current.

Welding Safety

Have you ever heard the saying "some people are accident-prone"? The implication is that accidents just seem to follow some individuals no matter what they do. They just seem plagued with bad luck. Actually, there is no such thing as being accident-prone. People have accidents simply because they are careless, or indifferent to safety regulations.

Each year thousands of people suffer the pain of injury because they have failed to use good judgment, Fig. 1-2. In many ways, safety can be considered a habit, a kind of behavior. A habit is acquired; you are not born with it. It is the result of repetition—doing something over and over again until it becomes part of you. Thus if you consistently follow good safety practices you subconsciously build within yourself a safety awareness that usually keeps you from making foolish mistakes. Taking a chance is another way of saying "I don't know any better." Thus safety simply means using a little common sense to avoid many serious accidents.

Fig. 1-2. Being injured because of some foolish accident is no fun.

Finally safety is not something you read about or practice only on occasion. It has to be observed constantly. Industry places a high premium on safety—ask anyone in industry and they will tell you that a tremendous amount of time and effort is given to safety. So never take chances; you will enjoy your work more if you learn to become a safe worker.

Accident reporting. Always report an accident regardless of how slight it may be. Even a little scratch might lead to an infection, or a minute particle could result in a serious eye injury. Prompt attention to any injury usually will minimize what may become serious if neglected.

Generally, where any physical work is performed, either in a learning situation or on an actual payroll job in industry, a definite accident reporting procedure is established. Since this reporting is for the best interest of the individual, it is foolhardy to ignore it or go around it. Consequently, make it a practice to become fully informed about what should be done and then take immediate action if an accident occurs.

Work behavior. On some occasion you may be tempted to engage in what might appear to be a harmless prank. Any form of horseplay in a shop is dangerous and can lead to an accident. There are many recorded incidents where foolish play ended in a serious injury. Most work areas are reasonably safe if proper work precautions are taken, but no one is safe if good work attitudes are ignored.

Welding equipment familiarization. No welding equipment of any kind should ever be used until exact instructions on how to operate it have been received. Manufacturer's recommended methods are very important and should be followed at all times. Attempting to operate a piece of equipment without instruction not only may damage the equipment but could result in a serious injury. Welding equipment of all kinds is safe to operate, providing it is used in the proper manner.

CAUTION: In operating welding equipment, never try to remedy a malfunction without first consulting supervising personnel. This applies to a whole range of items from a leaking gas hose to a loose cable on a welding generator.

4 Arc Welding

The instructor or foreman knows what should be done, and it is his responsibility to decide on the course of action to be taken.

Ventilation. All welding should be done in well ventilated areas. There must be sufficient movement of air to prevent accumulation of toxic fumes or possible oxygen deficiency. Adequate ventilation becomes extremely critical in confined spaces where dangerous fumes, smoke, and dust are likely to collect.

Types of Joints and Welds

The five basic joints used in welding are butt, T, lap, edge, and corner. See Fig. 1-3.

These various joints are used with the following types of welds: *surfacing*, *fillet*, and *groove*. See Fig. 1-4.

Surfacing weld. A type of weld composed of one or more stringer or weave beads deposited on an unbroken surface to obtain desired properties or dimensions.

Fillet weld. A fillet weld is approximately a triangle in cross-section, joining two surfaces at right angles to each other in a lap, T or corner joint.

Groove weld. A groove weld is a weld made in the groove between two members to be joined. The weld is adaptable for joints classified as square butt, single-V, double-V, single-U, double-U, single-J and double-J.

Welding Positions

The four positions assumed in welding are flat, overhead, vertical and horizontal. See Fig. 1-5. The flat position is the most widely used because welding can be done faster and easier. Overhead welding is somewhat more difficult, because the molten metal has a tendency to sag and considerable skill is required to secure a uniform bead with proper penetration. Vertical welding is done in a vertical line from the bottom to the top or top to bottom of the plates. On thin material a downhill welding technique is usually more applicable. Horizontal welding is also difficult to perform because, as in overhead welding, the molten puddle has a tendency to sag.

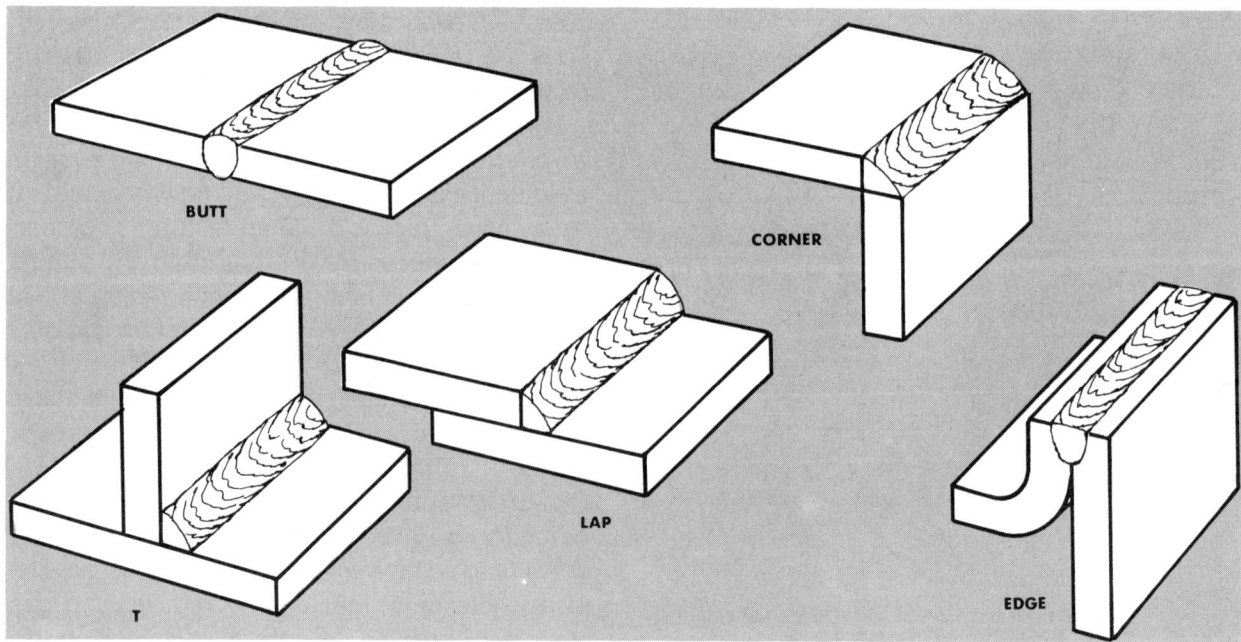

Fig. 1-3. Basic types of joints.

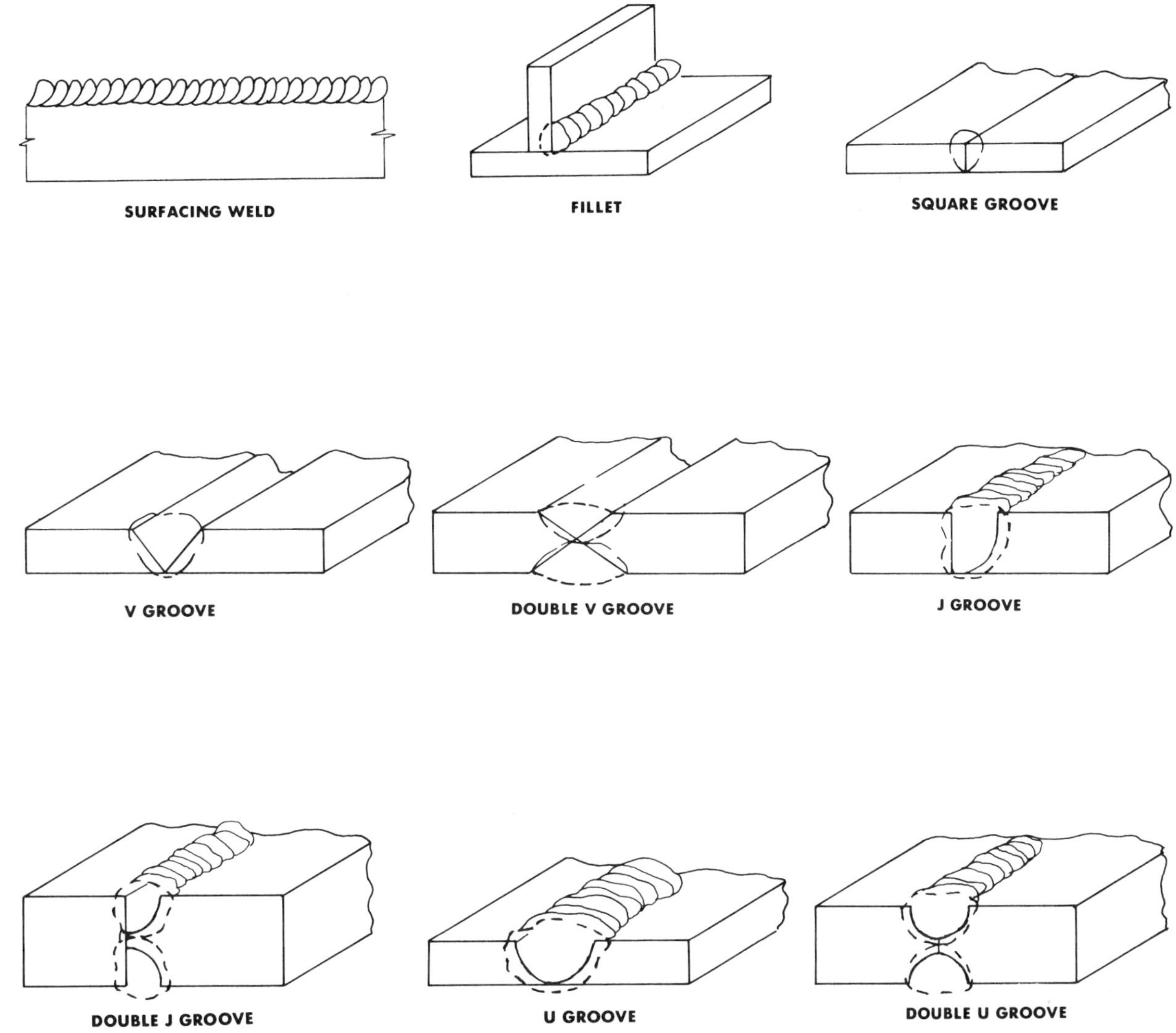

Introduction to Welding 5

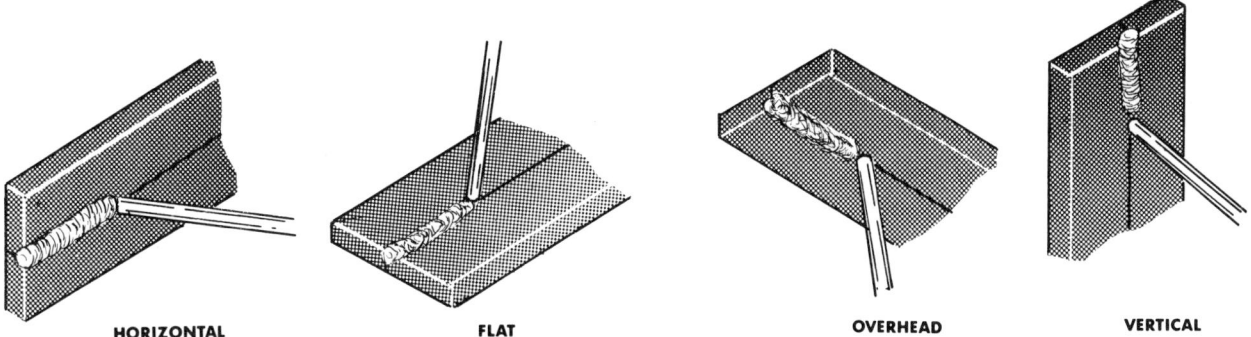

Fig. 1-4. Types of welds.

Fig. 1-5. There are four main welding positions.

Basic Welding Terms

Before proceeding with various welding operations, an understanding of the following terms is important:

Electrode. Thin metal wire, coated with a special substance and used as a filler to join the metal to be welded.

Base metal or parent metal. Metal to be welded. See Fig. 1-6.

Bead. Layer or layers of metal, as shown in Fig. 1-6, deposited on the base metal as the electrode melts.

Ripple. Shape of the deposited bead that is caused by movement of the electrode as shown in Fig. 1-6.

Pass. Each layer of beads deposited on the base metal as in Fig. 1-6.

Crater. Depression in the weld bead or molten pool, as shown in Fig. 1-6, made by the arc as the electrode comes in contact with the base metal.

Penetration. Depth of fusion with the base metal, shown in Fig. 1-7.

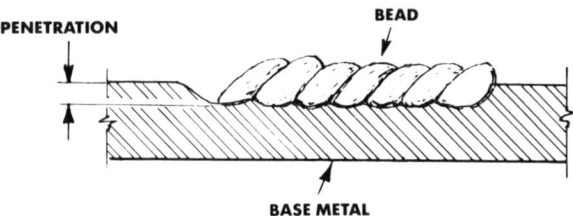

Fig. 1-7. The depth of the surface of the base metal to the bottom of the bead is called penetration.

Reinforcement. Refers to the amount of weld metal that is piled up above the surface of the pieces being joined. It is particularly applicable to butt welds. See Fig. 1-8.

Toes. The points where the base metal and weld metal meet. See Fig. 1-8.

Face. The exposed surface of the weld bounded by the toes of the weld. The face may be either concave or convex. See Fig. 1-8.

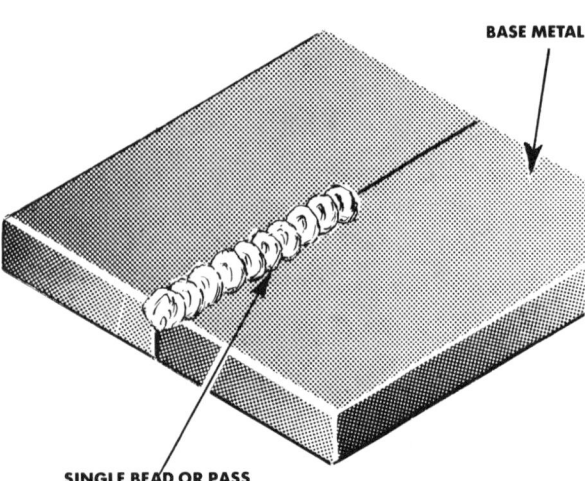

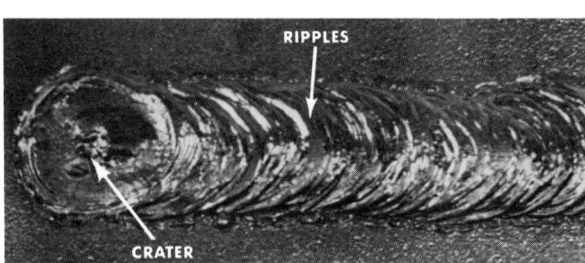

Fig. 1-6. These views illustrate bead, ripple, pass, and the base metal.

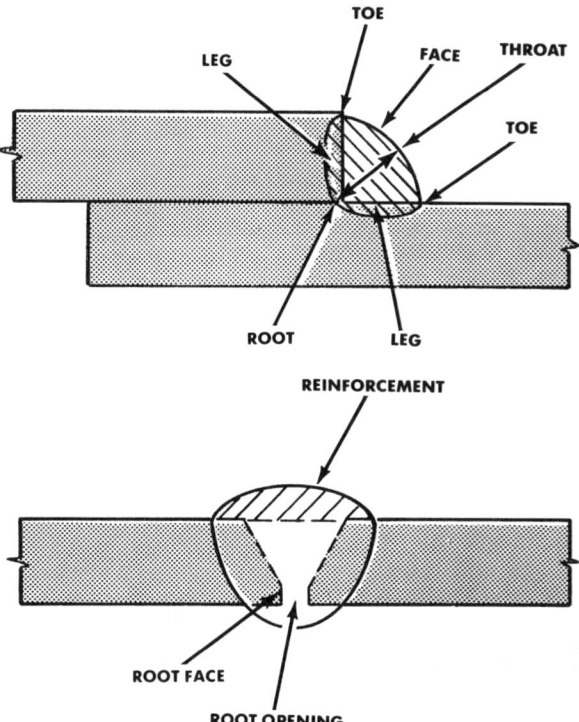

Fig. 1-8. Terms used in indicating the various parts of a weld, as defined in the text.

Root. The point of the weld triangle opposite the face of the weld. See Fig. 1-8.

Root face. The bottom lip near the slanted surface of the groove. See Fig. 1-8.

Throat. The distance through the center of the weld from the face to the root. See Fig. 1-8.

Effective throat. Size of a weld extending from the root of the weld to the top surface of the base metal. See Fig. 1-9.

Weld size. Same as effective throat. See Fig. 1-9.

Weld legs. The vertical and horizontal distances of a weld extending from the root of a joint to the toes of the weld. See Fig. 1-8.

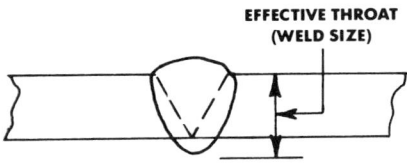

Fig. 1-9. Effective throat of a weld.

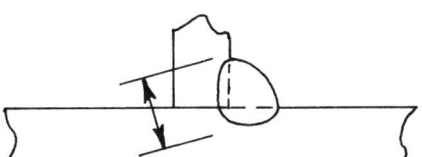

SELF QUIZ

Correct answers are listed in the back of the book.

True and False

Circle the letter T if the statement is true or the letter F if it is false.

1. T F Experienced welders are usually considered to be in one of the higher classification of job wages.
2. T F Welders do not need any understanding of basic mathematics or print reading.
3. T F A general maintenance welder needs only a limited background in just one area of welding.
4. T F Both gas metal-arc welding and gas tungsten-arc welding are done under a protective shield of gas.
5. T F In the gas tungsten-arc welding process the arc is formed between a consumable electrode and the work.
6. T F If you find something is wrong with your welding machine, you should fix it immediately yourself.

Short Answer

Complete each statement and place the answer in the blank.

7. A small deposit of weld metal on a plate is called a _____.
8. The depression in the base metal made by the arc of the electrode is known as a _____.
9. Each layer of beads deposited on the base metal is called a _____.
10. The shape of the deposited beads is referred to as _____.
11. A weld made on a lap joint is known as a _____.

Arc Welding

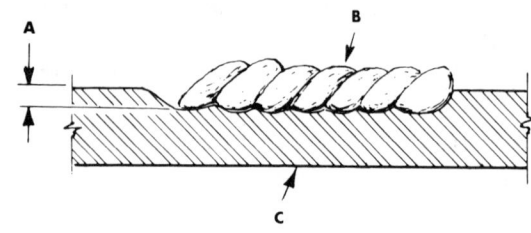

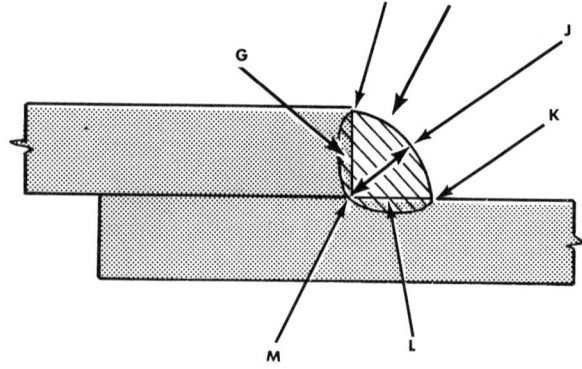

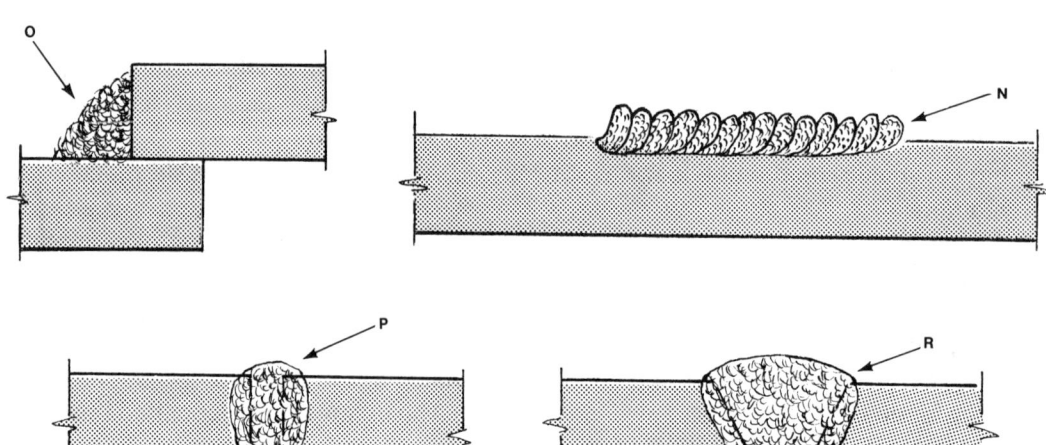

Identification

12. The illustrations on this page show different parts of a weld and types of joints. Identify each lettered part and write your answer in the appropriate blank.

A _____
B _____
C _____
D _____
E _____
F _____
G _____
H _____
I _____
J _____
K _____
L _____
M _____
N _____
O _____
P _____
R _____

UNIT 2
Reading Weld Symbols

In the fabrication of metal products, the welder usually has to work from a print which shows in detail exactly how the structure is to be made. He or she will find on the print not only where the welds are to be located, but also the type of joint to be used, as well as the correct size and amount of weld to be deposited at the designated seams. This information is indicated by a set of symbols which have been standardized by the American Welding Society (AWS).

Some of the more common symbols for weldments are included in this unit. For a more complete description see bulletin *AWS A2.4.*

Designating Types of Welds

The main foundation of the weld symbol is a reference line with an arrow at one end, as shown in Fig. 2-1. Notice that above or below the *base line* the type of weld is indicated—whether it is a fillet, groove, flange, plug, spot or seam weld. Also included is such information as surface contour of a weld, size of a weld bead, length of a weld, bead finish and often the type of welding process to be used. All of these data are indicated by geometric figures, numerical values, and abbreviations. Geometric figures,

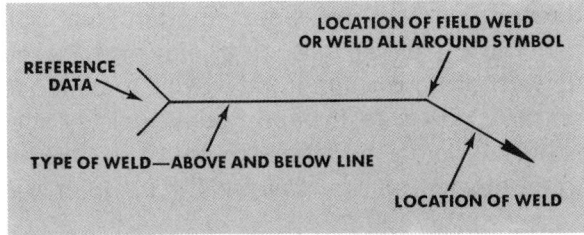

Fig. 2-1. Base of weld symbol.

10 Arc Welding

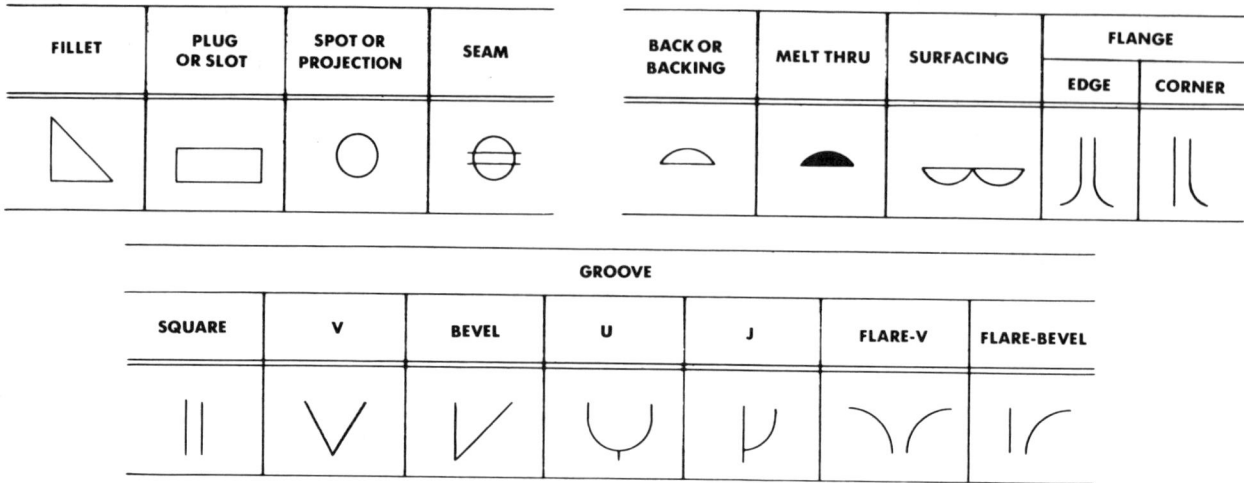

Fig. 2-2. Weld symbols (AWS)

such as seen in Fig. 2-2, are placed above or below the reference line in designating a weld.

Location of Symbols

A weld is said to be either on the *arrow* or *other* side of a joint. The arrow side is the surface that is in direct line of vision, while the other side is the opposite surface of the joint. See Fig. 2-3.

Weld location is designated by running the arrowhead of the reference line to the joint. The direction of the arrow is not important; that is, it can run on either side of a joint and extend upward or downward. If the weld is to be made on the *arrow side*, the appropriate weld symbol is placed *below* the reference line. If the weld is to be located on the *other side* of the joint, the weld symbol is placed *above* the reference line. When both sides of the joint are to be welded, the same weld symbol appears above and below the reference line. See Fig. 2-3.

Fillet, bevel, J-groove, flare-bevel groove, and corner-flange weld symbols are shown with the *perpendicular leg* always to the left of the weld symbol. See Fig. 2-4 for example.

The size of a weld is shown to the left of the weld symbol and the length to the right of the weld symbol. Sizes are expressed in decimal and fractional inches or in metric units. See Fig. 2-5.

Fillet welds. When both sides of a fillet are to be welded and both welds have the same dimensions, both are dimensioned. If the welds differ in dimensions, both are dimensioned. Where a note appears on a drawing that governs the size of a fillet weld, no dimensions are usually shown on the symbol.

When a fillet weld with unequal legs is required, the size of the legs is placed in parentheses as shown in Fig. 2-6.

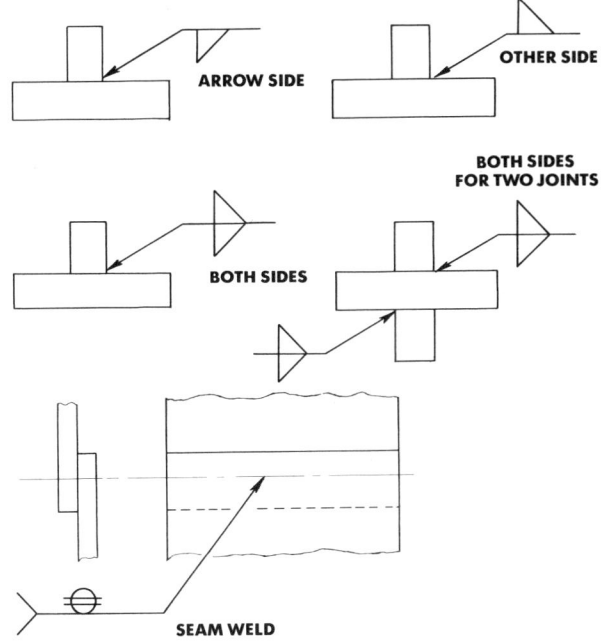

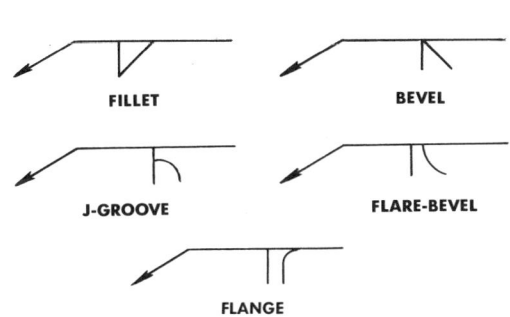

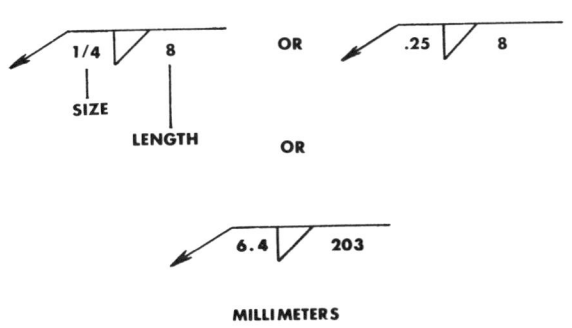

Fig. 2-3. How weld locations are designated. (AWS)

Fig. 2-4. The perpendicular leg of these weld symbols will be found on the left.

Fig. 2-5. Methods of expressing weld sizes.

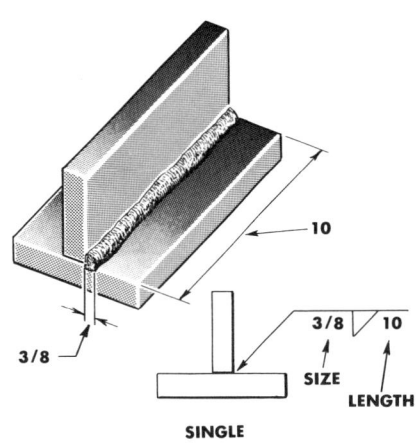

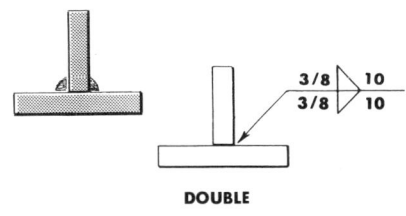

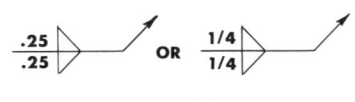

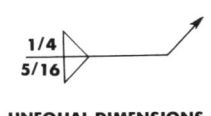

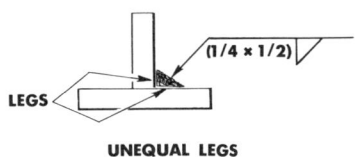

Fig. 2-6. How size and length of fillet welds are indicated.

Intermittent fillet welds. The length and pitch increments of intermittent welds are shown to the right of the weld symbol. The first figure represents the length of the weld section and the second figure the pitch (center-to-center spacing) between welds. See Fig. 2-7.

A staggered intermittent fillet weld is shown by staggered weld symbols above and below the reference line as shown in Fig. 2-7.

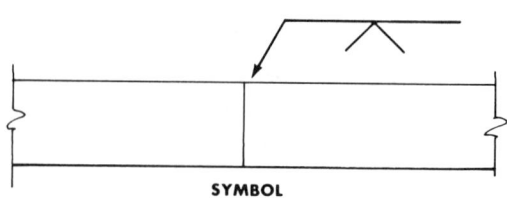

Fig. 2-8. Size is not shown for single and symmetrical double-groove welds with complete penetration.

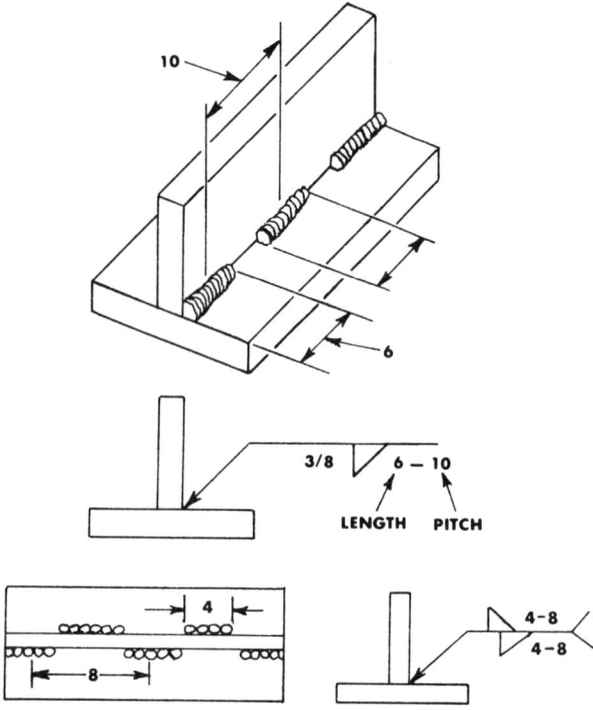

Fig. 2-7. How length and pitch of intermittent fillet welds are indicated.

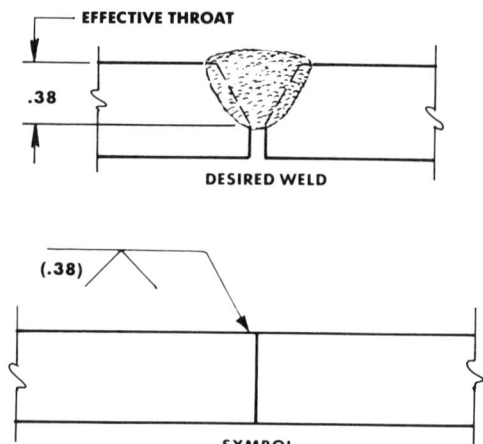

Fig. 2-9. How size is shown on grooved welds with partial penetration. (AWS)

Groove welds. There are several types of groove welds. Their sizes are shown as follows:

1. For those single-groove and symmetrical double-groove welds that extend completely through the members being joined, no size is included on the weld symbol. See Fig. 2-8.

2. For groove welds which extend only partly through members being joined or on non-symmetrical double-groove joints, weld size (effective throat) is shown in parentheses on the left of the weld symbol. See Fig. 2-9.

3. A dimension not in parentheses when placed to the left of the weld symbol indicates the depth of the bevel only. See Fig. 2-10. When both the effective throat and bevel size are indicated, the groove bevel depth is located to the left of the effective throat size as shown in Fig. 2-10.

4. An arrow with a definite break indicates which member of a groove joint is to be beveled. See Fig. 2-10.

5. Root opening of groove weld is shown inside the weld symbol. The included angle is placed below the weld symbol. See Fig. 2-11.

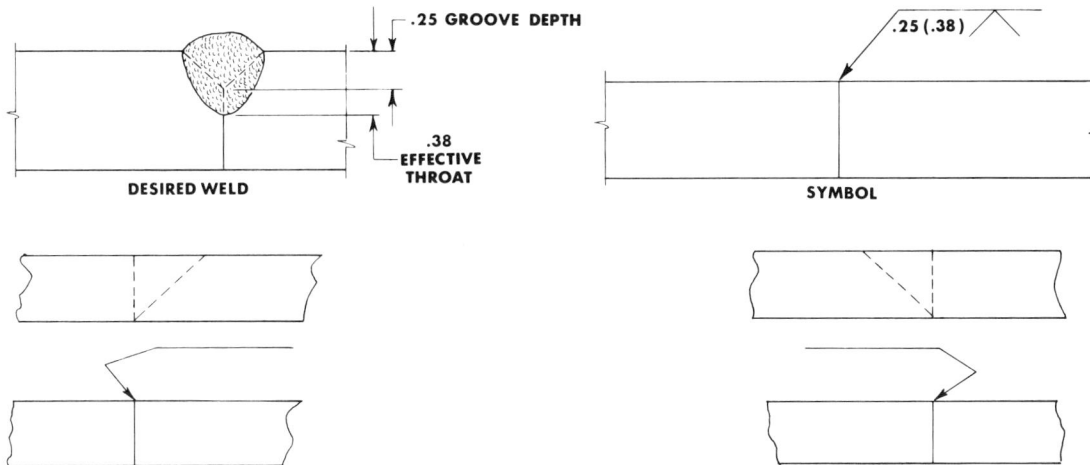

Fig. 2-10. How dimensions are used to show size and root penetration of grooved welds. (AWS)

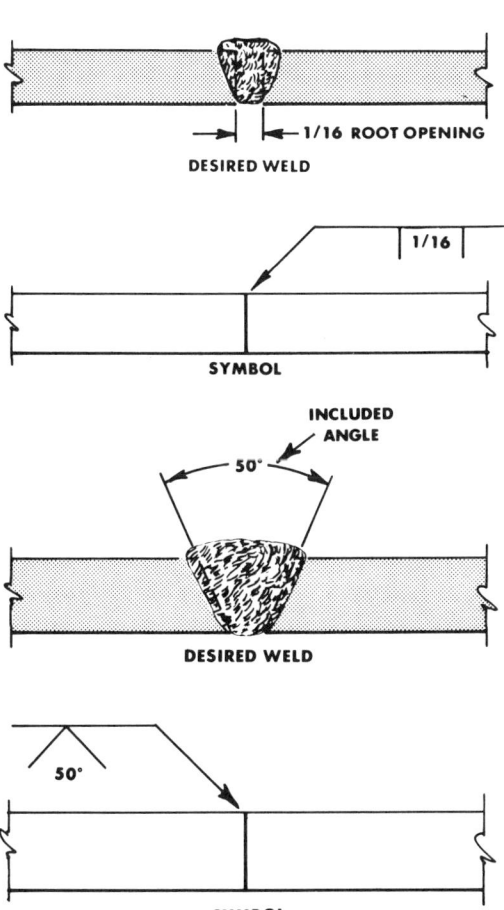

Fig. 2-11. How root opening and included angle are shown for groove welds. (AWS)

Flange welds. The radius and height of the flange are separated by a plus mark and placed to the left of the weld symbol. The size of the weld is shown by a dimension located outward of the flange dimensions. See Fig. 2-12.

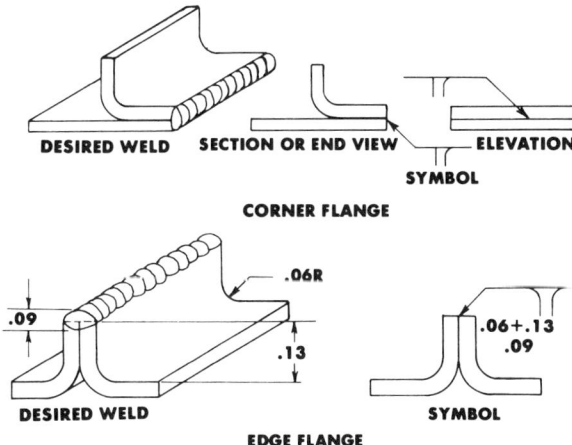

Fig. 2-12. How flange welds are indicated.

Combining Weld Symbols

In the fabrication of a product, there are occasions when more than one type of weld is to be made on a joint. Thus a joint may require both a fillet and double-bevel groove weld. When this happens, a symbol is shown for each weld. See Fig. 2-13.

14 Arc Welding

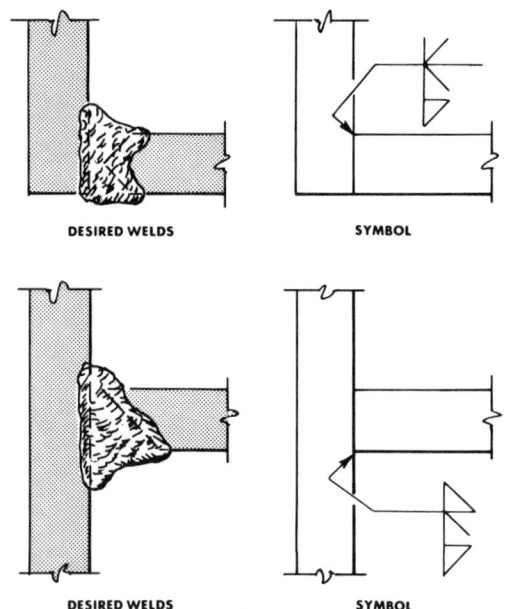

Fig. 2-13. Combined weld symbols.

Supplementary Symbols

Weld-all-around symbol. When a weld is to extend completely around a joint, a small circle is placed where the arrow connects the reference line. See Fig. 2-14.

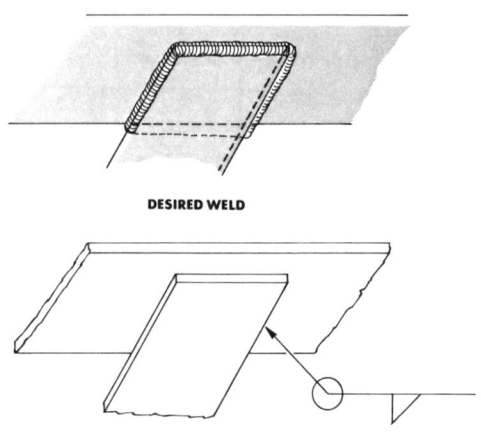

Fig. 2-14. Weld-all-around symbol. (AWS)

Field weld symbol. Welds that are to be made in the field (welds not made in a shop or at the place of initial construction) are indicated by a vertical flag. See Fig. 2-15.

Fig. 2-15. Field weld symbol. (AWS)

Reference tail. The tail is included only when some definite welding specification, procedure, reference, weld or cutting process needs to be called out; otherwise it is omitted. This data is often given in the form of symbols. See Fig. 2-16. Abbreviations in the tail may also call out some specifications which are included on some other part of the print.

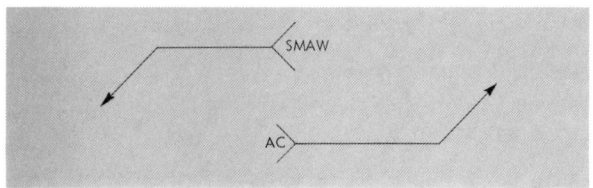

Fig. 2-16. The tail is used to indicate some specific detail or weld process. (AWS)

Surface contour of welds. When bead contour is important, a special flush, concave or convex contour symbol is added to the weld symbol. Welds that are to be mechanically finished also carry a finish symbol along with the contour symbols. See Fig. 2-17.

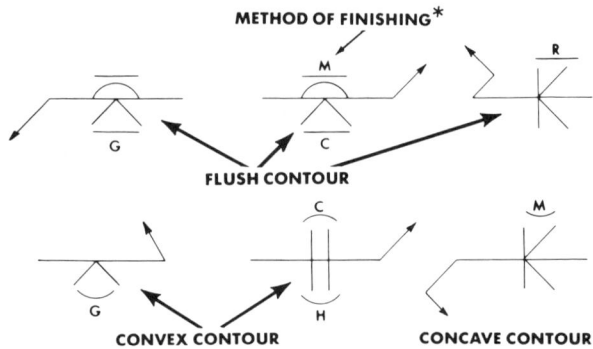

*Finish symbols used herein indicate the method of finishing ("C" = chipping; "G" = grinding; "M" = machining; "R" = rolling; "H" = hammering) and not the degree of finish.

Fig. 2-17. Method of showing surface contour of welds. (AWS)

Reading Weld Symbols **15**

Back or backing weld. Back or backing weld refers to the weld made on the opposite side of the regular weld. Back welds are occasionally specified to insure adequate penetration and provide additional strength to a joint. The particular symbol is included opposite the weld symbol. No dimensions of back or backing welds except height of reinforcement are shown on the weld symbol. See Fig. 2-18.

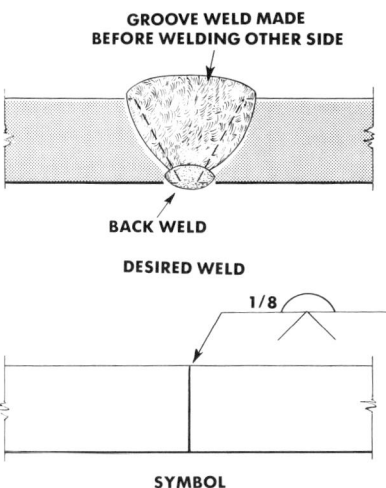

Fig. 2-18. Use of back weld symbol to indicate back weld. (AWS)

Melt-thru welds. When complete joint penetration of the weld through the material is required in welds made from one side only, a special melt-thru weld symbol is placed opposite the regular weld symbol. No dimension of melt-thru, except height of reinforcement, is shown on the weld symbol, as you will note in Fig. 2-19.

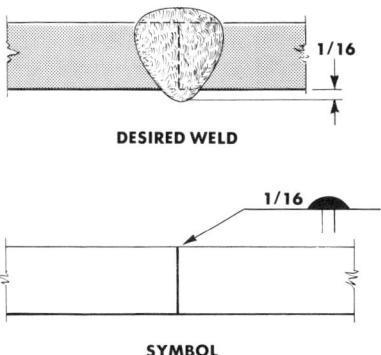

Fig. 2-19. Application of melt-thru symbol.

Surfacing welds. Welds whose surfaces must be built up by single or multiple pass welding are provided with a surfacing weld symbol. The height of the built-up surface is indicated by a dimension placed to the left of the surfacing symbol. See Fig. 2-20. The extent, location, and orientation of the area to be built up are normally indicated on the drawing.

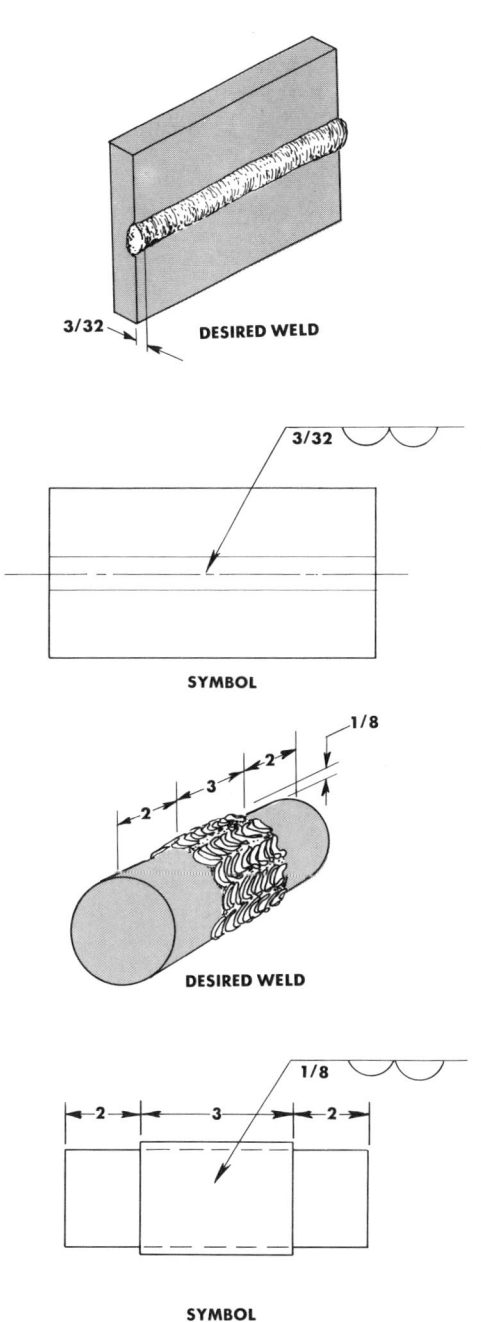

Fig. 2-20. Application of surfacing symbol to indicate surfaces built up by welding.

SELF QUIZ

Correct answers are listed in the back of the book.

Identification

Identify the weld symbols shown below and place the correct answers on the blanks.

1. What kind of joints do the following symbols represent?

A _____
B _____
C _____
D _____
E _____
F _____
G _____
H _____
I _____
J _____

2. What welding conditions do the symbols below specify?

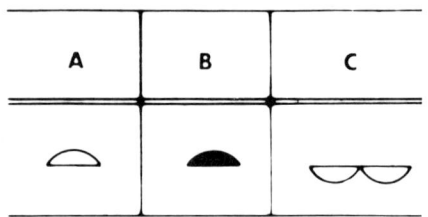

A _____
B _____
C _____

3. What do the following symbols mean?

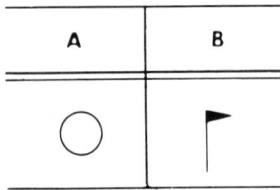

A _____
B _____
C _____
D _____
E _____

4. Identify the location of the following fillet welds.

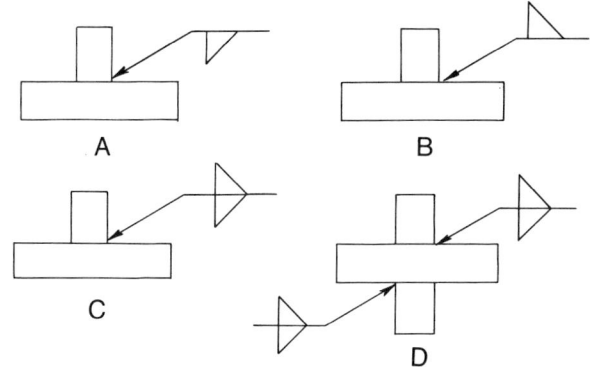

A _____
B _____
C _____
D _____

5. Describe how the weld shown below is to be made.

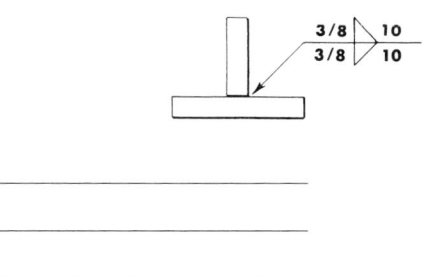

6. What do the following welds require?

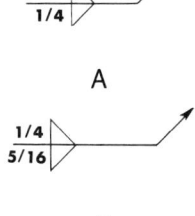

A _____

B _____

7. What kind of weld does the symbol below specify?

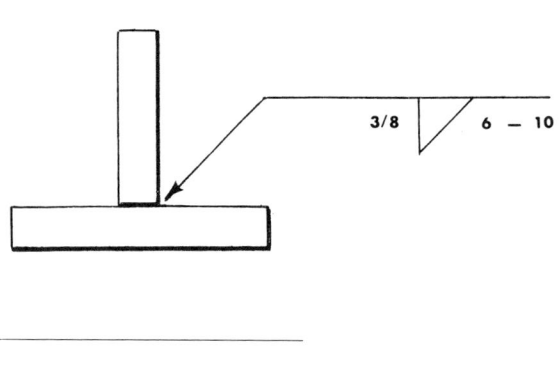

8. What does the following weld symbol mean?

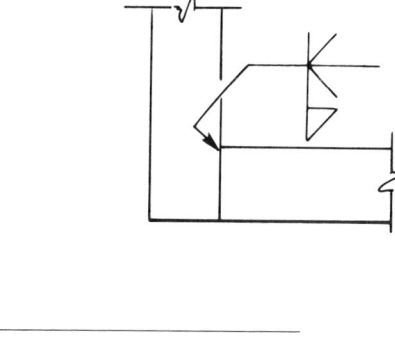

9. What do the numbers on the symbol below tell you?

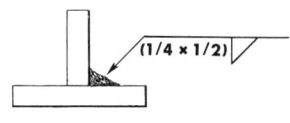

18 Arc Welding

10. What is required in joints A and B below?

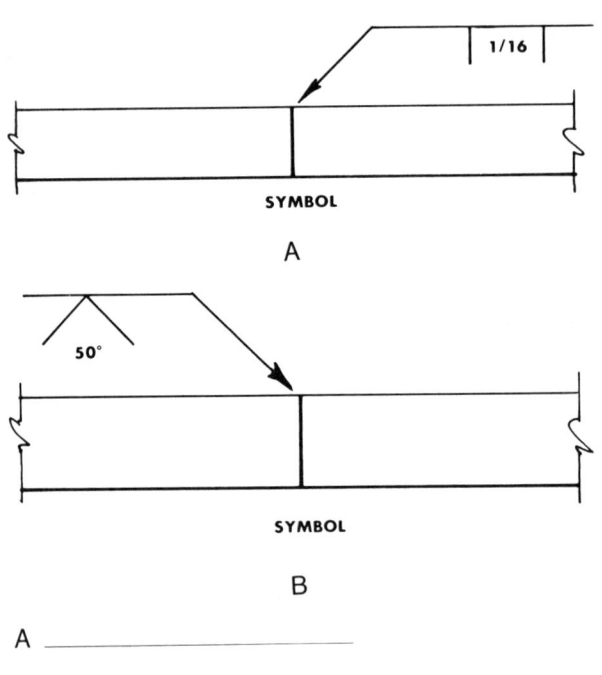

A _____
B _____

11. What does the symbol below mean?

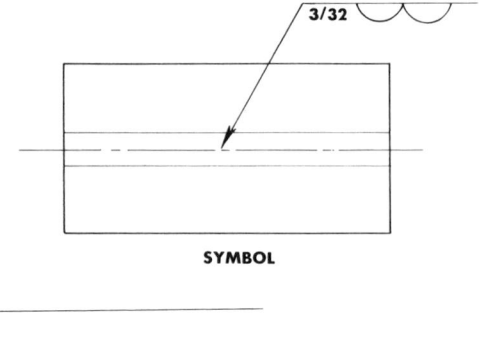

_____
_____
_____

12. What does the symbol below specify?

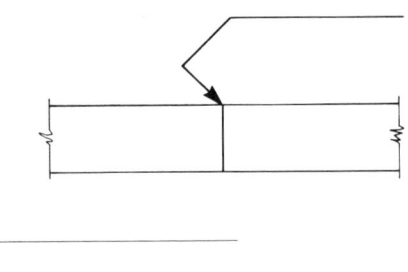

_____
_____

13. What do the parts in the following symbol represent?

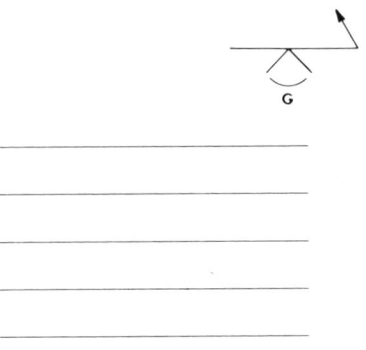

_____
_____
_____
_____
_____

14. What is the meaning of the following symbol?

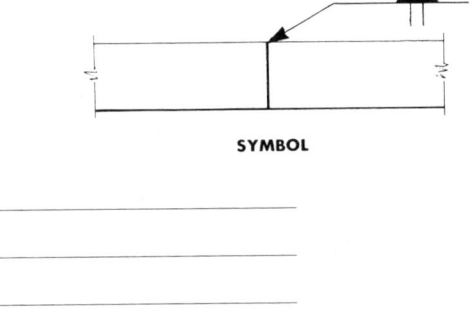

_____
_____
_____

15. Identify this weld symbol.

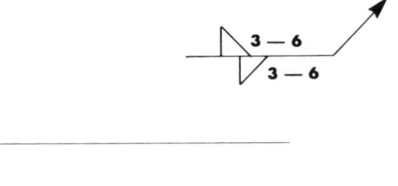

_____
_____

16. What does this arrow show?

_____

UNIT 3
Basic Welding Metallurgy

In preparing to become a skillful welder you should become familiar with the effects of heat on the structure of metal.

You will also need to know what safeguards must be followed in welding metals, because application of heat during the welding process may destroy the very elements which were originally added to improve the structure of the metal. For example, metals expand and contract, thereby setting up great stresses which often result in severe distortions. Improper welding of stainless steel may result in a complete loss of its corrosion-resistant qualities, and welding high-carbon steel in the same manner as low-carbon steel may produce such a brittle weld as to make the welded piece unusable.

MECHANICAL PROPERTIES OF METALS

Mechanical properties indicate how materials behave under applied loads. That is, they show how strong a metal is when it comes in contact with one or more forces.

Some of the basic terms that are associated with mechanical properties of metals are included in the paragraphs that follow. A welder should become familiar with them because they are often directly related to his ability to produce sound welds.

Stress is the internal resistance a material offers to being deformed and is measured in terms of the applied load over the area. See Fig. 3-1A.

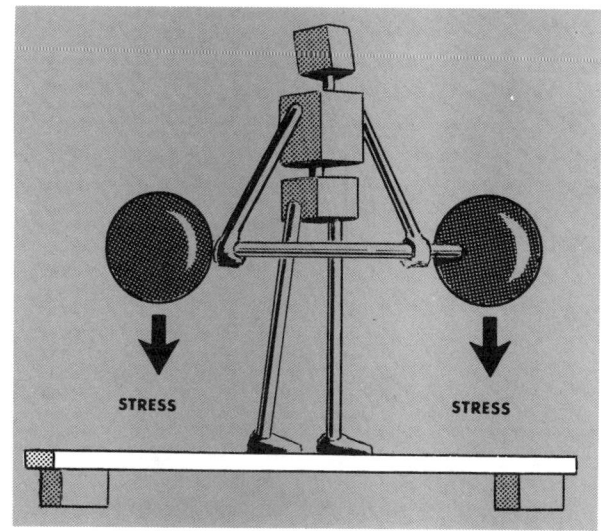

Fig. 3-1A. Example of stress.

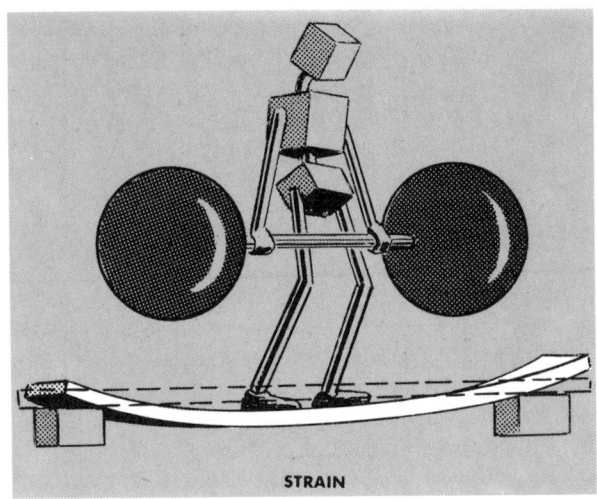

Fig. 3-1B. Example of strain.

Strain is the deformation that results from a stress and is expressed in terms of the amount of deformation per inch. See Fig. 3-1B.

Elasticity is the ability of a metal to return to its original shape after being elongated or distorted, when the forces are released. See Fig. 3-2. A rubber band is a good example of what is meant by elasticity. If the rubber is stretched, it will return to its original shape after you let it go. However, if the rubber is pulled beyond a certain point, it will break. Metals with elastic properties react in the same way.

Tensile strength is that property which enables the metal to resist forces acting to pull it apart. See Fig. 3-3. It is one of the more important factors in the evaluation of a metal.

Compressive strength is the ability of a material to resist being crushed. See Fig. 3-4. Compression is the opposite of tension with respect

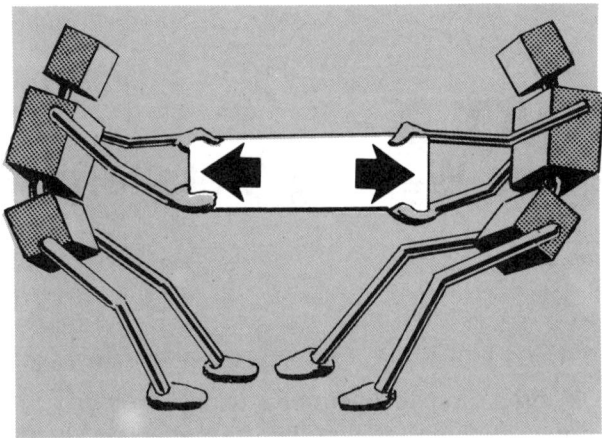

Fig. 3-3. A metal with tensile strength resists pulling forces.

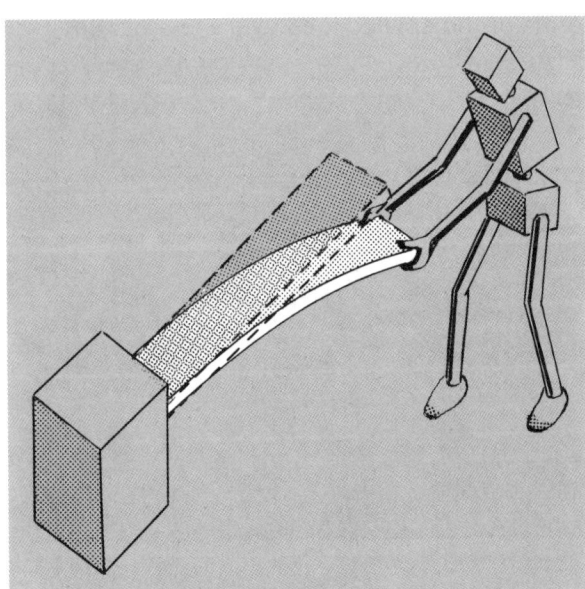

Fig. 3-2. A metal having elastic properties returns to its original shape after the load is removed.

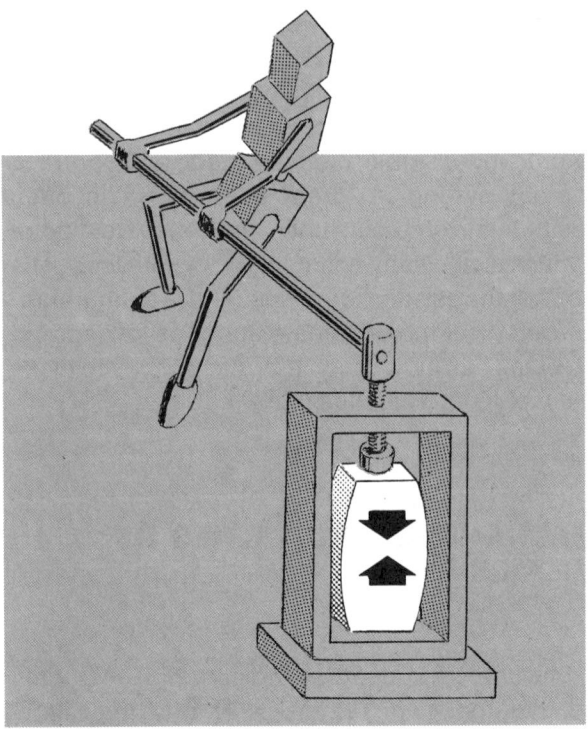

Fig. 3-4. Compressive strength refers to the property of metal to resist crushing forces.

Basic Welding Metallurgy 21

to the direction of the applied load. Most metals have high tensile strength and high compressive strength. However, brittle materials such as cast iron have high compressive strength but only moderate tensile strength.

Torsional strength is the ability of a metal to withstand forces that cause a member to twist. See Fig. 3-5, middle.

Shear strength refers to how well a member can withstand two equal forces acting in opposite directions. See Fig. 3-5, bottom.

Bending strength is that quality which resists forces from causing a member to bend or deflect in the direction in which the load is applied. Actually a bending stress is a combination of tensile and compressive stresses. See Fig. 3-5, top, to grasp the idea.

Impact strength is the ability of a metal to resist loads that are applied suddenly and often at high velocity. The higher the impact strength of a metal the greater the energy required to break it. Impact strength may be seriously affected by welding, since it is one of the most structure sensitive properties.

Ductility refers to the ability of metal to stretch, bend, or twist without breaking or cracking. See Fig. 3-6. A metal having high ductility, such as

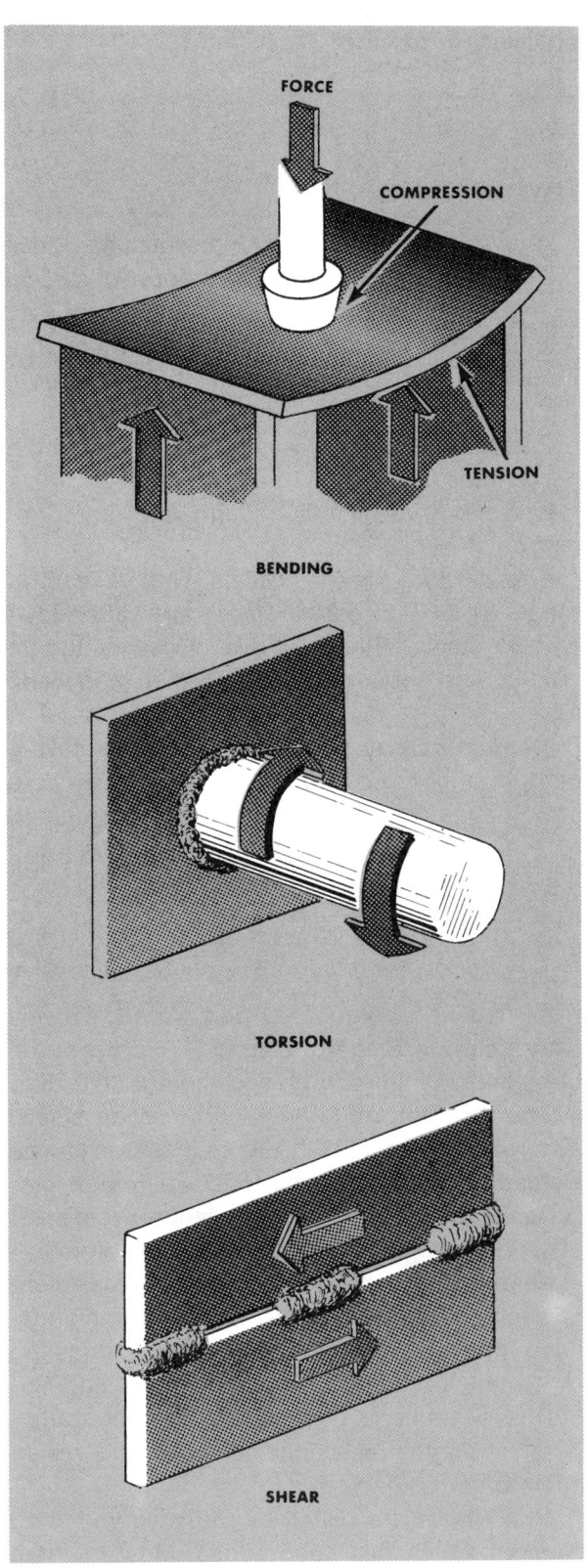

Fig. 3-5. Examples of bending, torsion, and of shearing stresses.

Fig. 3-6. A ductile metal can easily be shaped.

copper or soft iron, will fail or break gradually as the load on it is increased. A metal of low ductility, such as cast iron, fails suddenly by cracking when subjected to a heavy load.

Hardness is the ability of steel to resist indentation or penetration. See Fig. 3-7. Hardness is usually expressed in terms of the area of an indentation made by a special ball under a standard load, or the depth of a special indenter under a specific load.

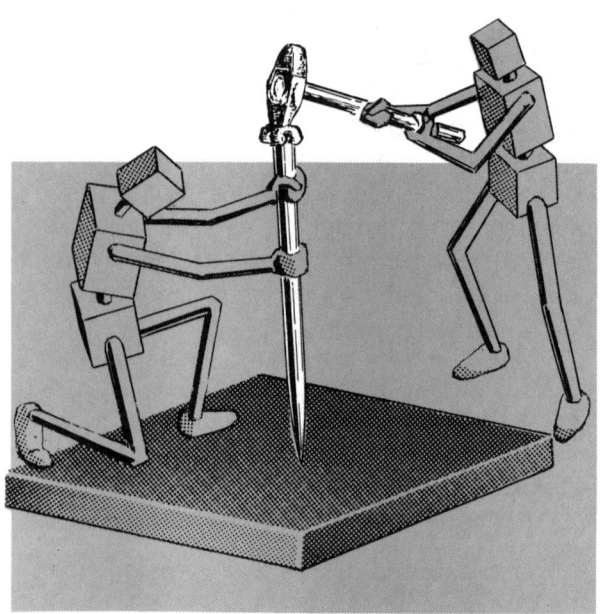

Fig. 3-7. Hardness resists penetration.

Brittleness is a condition whereby a metal will easily fracture under low stress. It is a property which often develops because of improper welding techniques. Brittleness is a complete lack of ductility.

Toughness may be considered as strength, together with ductility. A tough material or weld is one which may absorb large amounts of energy without breaking.

Malleability is the ability of a metal to be deformed by compression forces without developing defects, such as encountered in rolling, pressing, or forging.

Types of Steels

A plain carbon steel is one in which carbon is the only alloying element. The amount of carbon in the steel controls its hardness, strength, and ductility. The higher the carbon content, the harder the steel. Conversely, the less the carbon, the greater the ductility of the steel.

Carbon steels are classified according to the percentage of carbon they contain. They are referred to as low, medium, high, and very-high-carbon steels.

Low-carbon steels. Steels with a carbon range of 0.05 to 0.30 percent are called low-carbon steels. Steels in this class are tough, ductile, and easily machined, formed, and welded.

Medium-carbon steels. These steels have a carbon range from 0.30 to 0.45 percent. They are strong and hard but cannot be worked or welded as easily as low-carbon steels. Successful welding of these steels often requires special electrodes, but even then great care must be taken to prevent formation of cracks around the weld area.

High and very-high-carbon steels. Steels with a carbon range of 0.45 to 0.75 percent are classified as high-carbon and those with 0.75 to 1.7 percent carbon as very-high-carbon steels. As a rule, steels up to 0.65 percent carbon can be welded with special electrodes, although preheating and stress relieving techniques must often be used after the welding is completed.

An alloy steel is a steel to which one or more of such elements as nickel, chromium, manganese, molybdenum, titanium, cobalt, tungsten, or vanadium have been added. The addition of these elements gives steel greater toughness, strength, resistance to wear, and resistance to corrosion.

Alloy steels are called by the predominating element which has been added. Most of them can be welded, provided special electrodes are used.

Controlling Expansion and Contraction Forces

The strength of a welded joint depends a great deal on the way you control the expansion and contraction of the metal during the welding operation. Whenever heat is applied to a piece of metal, expansion forces are created which tend to change the dimensions of the piece. Upon cooling, the metal undergoes a change again as it attempts to resume its original shape.

No serious consideration is given these factors when there are no restricting forces to prevent the free movements of the expansion and contraction forces or when welding ductile metal, because the flow of metal will usually relieve the stresses. When free movement is restricted, a warping or distortion is likely to occur if the metal is malleable or ductile, and a fracture if the metal is brittle, as with cast iron.

The following are a few simple procedures which will help control the forces caused by expansion and contraction:

Proper edge preparation and fit-up. Make certain that the edges are correctly beveled. Proper edge beveling will not only restrict the effects of distortion but will insure good weld penetration. See Fig. 3-8. Although sometimes the bevel angle can be reduced, care must be taken to insure that there is sufficient room in the joint to permit proper manipulation of the electrode when doing the weld.

On long seams, especially on thin sections, the practice is to allow about $1/8''$ [0.125''] at the end of the weld for each foot in length for expansion. See Fig. 3-9 for example.

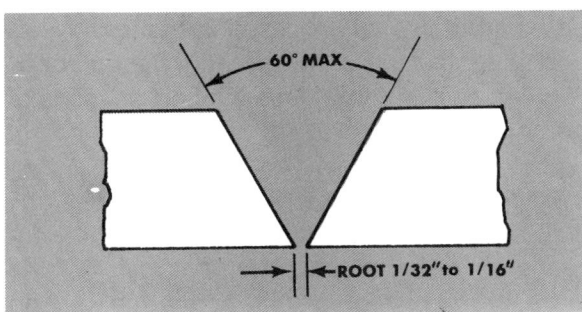
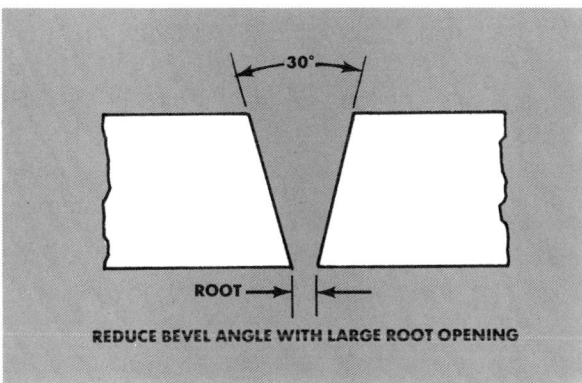
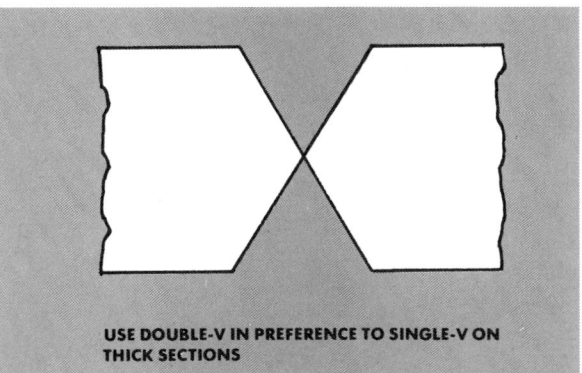

Fig. 3-8. Proper edge preparation will minimize distortion.

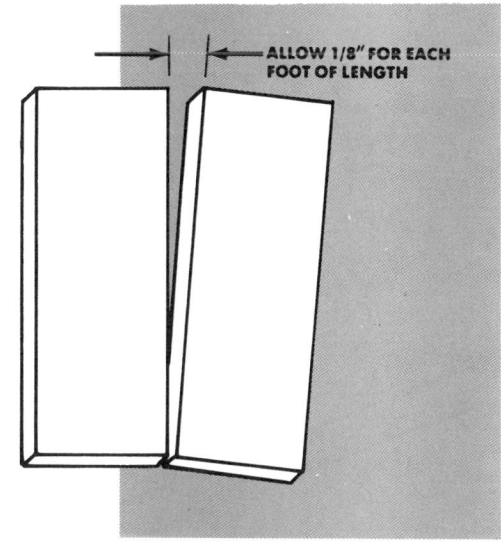

Fig. 3-9. Provide a space between the edges to be welded.

24 Arc Welding

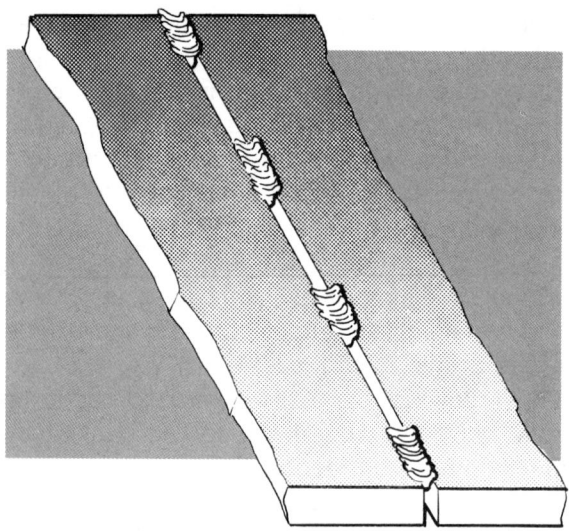

Fig. 3-10. Tacking the plates will hold them in position.

Tack welds are also used to control expansion on long seams as shown in Fig. 3-10. Tack welds are spaced about 12" apart and run approximately twice as long as the thickness of the weld. When tack welds are used, progressive spacing is not necessary. The plates are simply spaced an equal distance throughout the seam.

Minimizing heat input. Controlling the amount of heat input is somewhat more difficult for the beginner. An experienced welder is able to join a seam with the minimum amount of heat by rapid welding.

A technique often employed to minimize the heat input is the *intermittent,* or *skip weld.* Instead of making one continuous weld, a short weld is made at the beginning of the joint. Next, a length of a few inches is welded at the center of the seam, and then a short length at the end of the joint. Finally you return to where the first weld ended and proceed in the same manner, repeating the cycle until the weld is completed. See Fig. 3-11.

The use of the *back-step,* or *step-back,* welding method also minimizes distortion. With this technique, instead of laying a continuous bead from left to right, you deposit short sections of the beads from right to left as illustrated in Fig. 3-12.

Preheating. On many pieces, particularly alloy steels and cast iron, expansion and contraction forces can be better controlled if the entire structure is preheated before the welding is started. To be effective, preheating must be kept uniform throughout the welding operation, after the weld is completed the piece must be allowed to cool slowly. Preheating can be done with an oxy-acetylene or carbon flame.

Peening. To help a welded joint stretch as it cools, a common practice is to peen it lightly with the round end of a ball peen hammer.

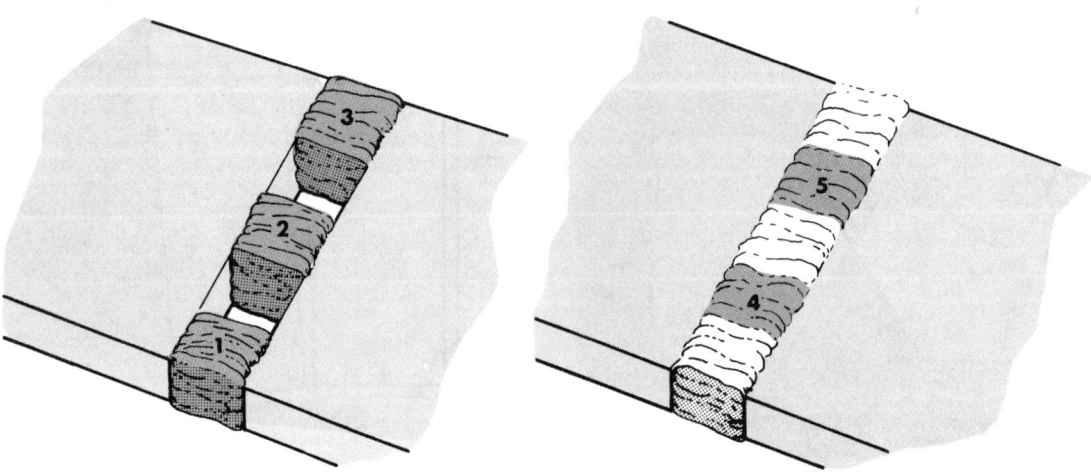

Fig. 3-11. The intermittent weld, sometimes referred to as the skip weld, will prevent distortion.

Basic Welding Metallurgy 25

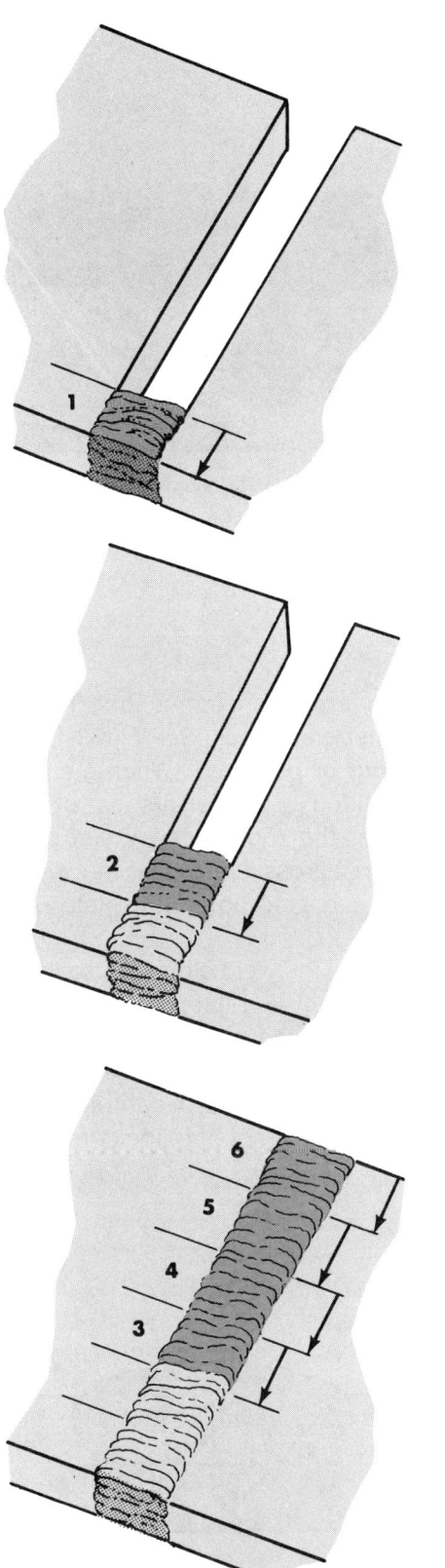

Fig. 3-12. This is how the back-step welding technique is done.

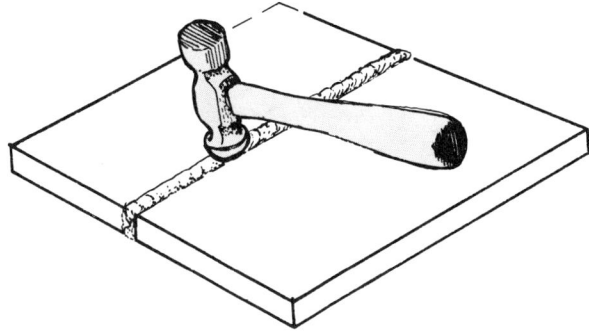

Fig. 3-13. Peening a weld helps to release the locked-up stresses.

However, peening should be done with care, because too much hammering will add stresses to the weld or cause the weld to work-harden and become brittle. See Fig. 3-13.

Jigs and fixtures. The use of jigs and fixtures will help prevent distortion, since holding the metal in a fixed position prevents excessive movements. A jig or a fixture is any device that holds the metal rigidly in position during the welding operation. Fig. 3-14 illustrates a simple way to hold pieces firmly in a flat position. These

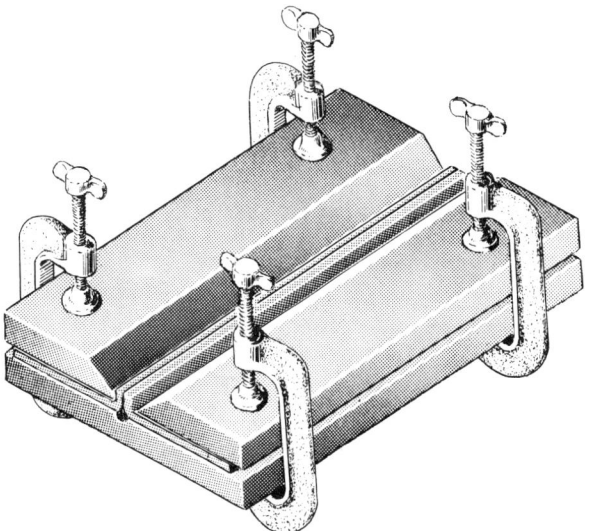

Fig. 3-14. Clamping pieces between heavy blocks will keep them straight.

26 Arc Welding

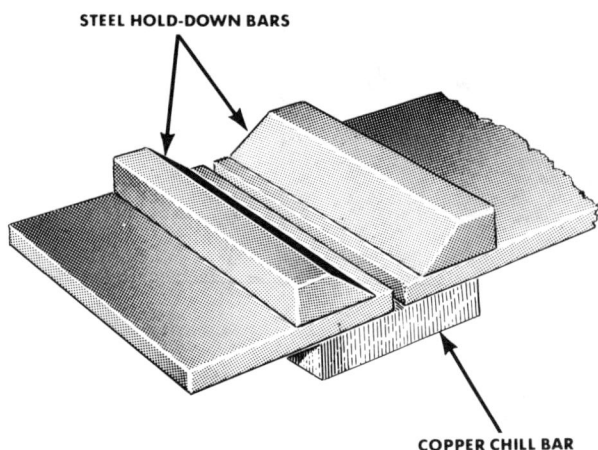

Fig. 3-15. Chill plates help to reduce heat and warpage in the weld area. (Republic Steel Corp.)

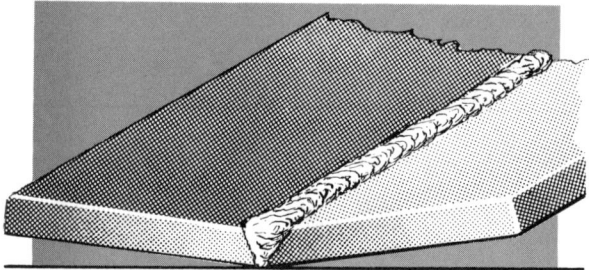

Fig. 3-17. Unless the plates are clamped, a butt joint will be distorted when welded.

heavy plates not only prevent distortion but they also serve as *chill blocks* to avoid excessive heat building up in the work. Special chill plates made of copper or other metal having good conductivity are particularly effective in dissipating heat away from the weld area. See Fig. 3-15.

Jigs and fixtures are used extensively in production welding since they permit greater welding speed while reducing to a minimum any form of distortion. By and large, industrial jigs and fixtures are designed to accommodate the specific production work being done.

Number of passes. Distortion can be kept to a minimum by using as few passes as possible over the seam. Two passes made with large electrodes are often better than three or four with smaller electrodes. See Fig. 3-16.

Parts out of position. When a single-V butt joint is welded, a greater amount of heat at the top than at the root of the V will cause more contraction across the top of the welded joint. The result is a distortion of the plate as shown in Fig. 3-17.

In a T-joint, the weld along the seam will bend both the upright and flat piece. See Fig. 3-18.

To minimize these distortions, the simplest thing to do is to angle the pieces slightly in the opposite direction in which contraction is to take place. Then, upon cooling, the contraction forces will pull the pieces back into position.

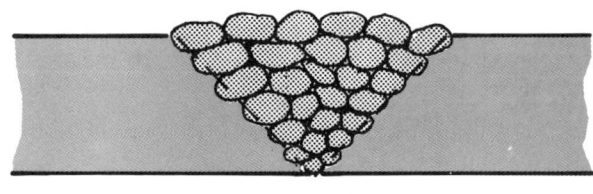

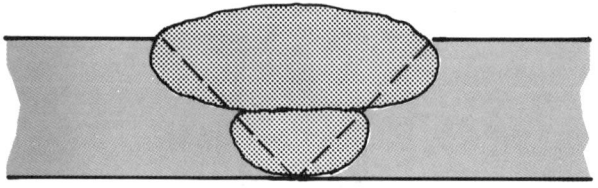

Fig. 3-16. Use few passes to reduce distortion.

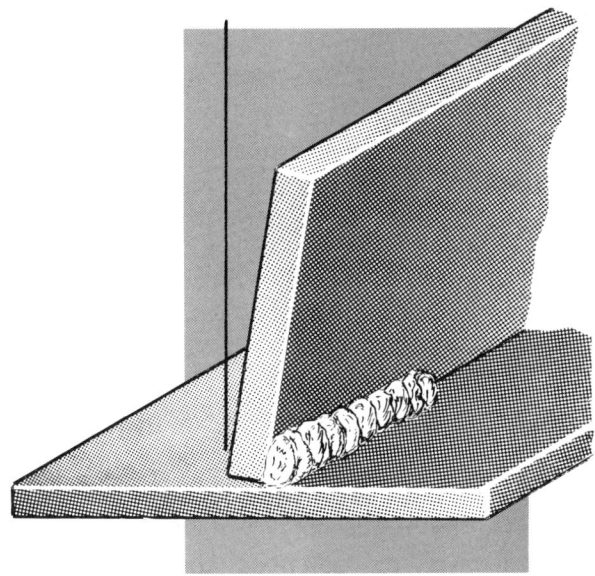

Fig. 3-18. A T-joint is likely to be distorted in this manner.

Welding Defects

In the process of welding various materials, precautions must be taken to prevent the development of certain defects in the weld metal; otherwise these defects will severely weaken the weld. The following are some of the principal defects that are significant in any welding or brazing process.

Grain growth. A wide temperature differential will exist between the molten metal of the actual weld and the edges of the heat-affected zone of the base metal. This temperature may range from a point far above the critical temperature down to a point which will not affect the metal. Thus the grain size can be expected to be large at the molten zone of the weld puddle and gradually reducing in size until recrystallization is reached. Grain growth can be kept to a minimum by effective control of preheating and postheating.

Where heavy sections require successive passes, it is possible to use the heat of each successive pass to refine the grain of the previous pass. This can be done only if the metal is allowed to cool below the lower critical temperature between each pass. High-carbon and alloy steels are especially vulnerable to coarse growth if cooled rapidly. These metals usually must be preheated before welding and then allowed to cool slowly after the weld is completed.

Blowholes. Blowholes are cavities caused by gas entrapment during the solidification of the weld metal. They usually develop because of improper manipulation of the electrode and failure to maintain the molten pool long enough to float out the entrapped gas, slag, and other foreign matter. When gas and other matter become trapped in the grains of the solid metal, small holes are left in the weld after the metal cools.

Blowholes can be avoided by keeping the molten pool at a uniform temperature throughout the welding operation. This can be done by using a constant welding speed so the metal solidifies evenly. Blowholes are most likely to occur during the stopping and starting of the weld along the seam, especially when the electrode must be changed.

Inclusions. Inclusions are impurities or foreign substances which are forced in a molten puddle during the welding process. Any inclusion tends to weaken a weld, because it has the same effects as a crack. A typical example of an inclusion is slag which normally forms over a deposited weld. If the electrode is not manipulated correctly, the force of the arc causes some of the slag particles to be blown into the molten pool. When the molten metal freezes before these inclusions can float to the top, they become lodged in the metal, producing a defective weld.

Inclusions are more likely to occur in overhead welding, since the tendency is not to keep the molten pool too long to prevent it from dripping off the seam. However, if the electrode is manipulated correctly and the right electrodes are used with proper current settings, inclusions can be avoided, or at least kept to a minimum.

Porosity. Porosity refers to the formation of tiny pinholes generated by atmospheric contamination. Some metals have a high affinity for oxygen and nitrogen when in a molten state. Unless an adequate protective shield is provided over the molten metal, gas will enter the metal and weaken it.

Crater cracks. These cracks occur across the weld bead crater and result from shrinkage. See Fig. 3-19. When a weld bead is deposited, solidification of the molten metal takes place from the sides and moves towards the center. The center of the crater cools rapidly because of the smaller amount of metal while the remaining bead cools more slowly, causing a concentration of stresses which eventually result in cracks. Crater cracks are particularly prevalent in thin concave fillet welds. See Fig. 3-20. Most crater cracks can be prevented by proper electrode manipulation. Care should be taken to completely fill up the craters and to round the weld bead slightly by using a shorter arc.

Root cracks. In making a groove or fillet weld, the first pass is in the form of a narrow stringer bead along the weld seam. This is followed by one or more layers of weld beads. It is the first layer, or root bead, that is the most susceptible to cracking. See Fig. 3-21. The cracking is generally due to the excessive carbon which the bead picks up from the base metal, thereby making the weld metal hard and brittle. As the weld metal cools, it shrinks, and as additional layers of beads are deposited on the root weld, tensile stresses form which develop into cracks. Root cracks can be prevented by preheating the base metal, using a more ductile weld metal, and providing sufficient space between the plates so they can move as the weld cools.

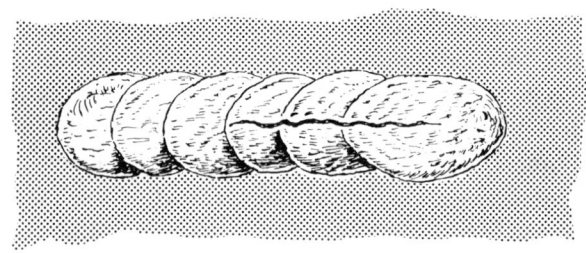
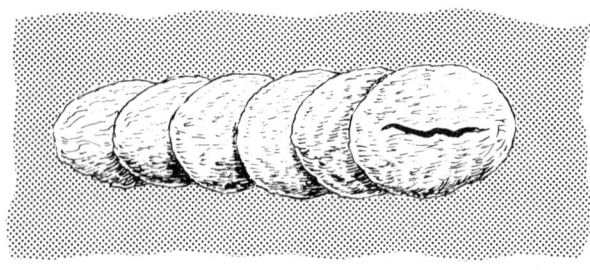

Fig. 3-19. Types of crater cracks.

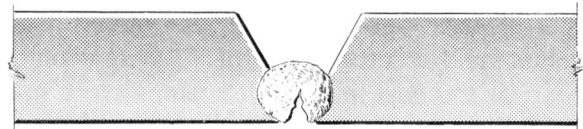

Fig. 3-21. The root bead is often more susceptible to cracking than later beads.

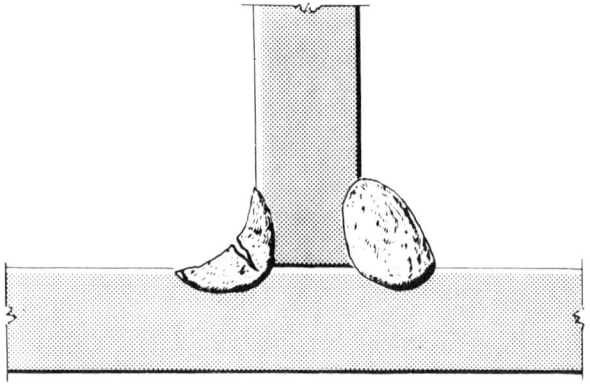

Fig. 3-20. Crater cracks will often occur on thin concave fillet welds.

Cold lap. Cold laps usually occur when the arc does not melt the base metal sufficiently, causing the slightly molten puddle to flow into the unwelded base metal. Very often if the puddle is allowed to become too large, this too will result in cold laps. For proper fusion, the arc should be kept at the leading edge of the puddle. When directed in such a manner, the molten puddle is prevented from flowing ahead of the welding arc. Also remember that the size of the puddle can be reduced by increasing the travel speed or reducing the amperage.

Insufficient penetration. Lack of penetration is due to a low heat input in the weld area or failure to keep the arc properly located on the leading edge of the puddle. If the heat input is too slow, increase the amperage.

Excessive penetration. Too much penetration or *burn through* is caused by having excessive heat in the weld zone. By reducing the amperage there will be less heat. Excessive penetration can also be avoided by increasing the travel speed. If the root opening in the joint is too wide, too much burn through may result. Usually improper design can be remedied by weaving the electrode.

SELF QUIZ

Correct answers are listed in the back of the book.

Multiple Choice

Circle the letter which represents the correct answer.

1. Which of these factors is the most important in any welding situation?
    a. alignment of the joint
    b. ductility of a metal
    c. protecting the weld from oxygen contamination
    d. allowing a weld to cool too slowly
2. Which of these elements is considered to be most harmful to metals?
    a. cold atmospheric temperatures
    b. corrosion
    c. high atmospheric temperatures
    d. paint
3. What is likely to occur if care is not taken when trying to relieve stresses in a weld by peening?
    a. There will be too much distortion.
    b. The weld will crack.
    c. The weld will become brittle.
    d. The bead shape will be ruined.
4. If a weld in high carbon steel is allowed to cool too rapidly, it
    a. becomes very ductile
    b. becomes very brittle
    c. cannot be annealed
    d. cannot be tempered
5. A steel with a carbon content ranging between 0.05 to 0.30 percent is classified as
    a. high carbon
    b. medium carbon
    c. low carbon
    d. very high carbon
6. Grain growth during a welding operation can be kept to a minimum by
    a. using high amperage settings
    b. peening the metal
    c. effective control of preheating and postheating
    d. rapid cooling
7. Blowholes are caused by
    a. improper electrode manipulation
    b. too high a current
    c. too low a current
    d. wrong type of electrode
8. Slag particles entrapped in the puddle are called
    a. residual stresses
    b. segregations
    c. air pockets
    d. inclusions

9. Stresses result from
   a. using too many passes
   b. improper provisions for expansion and contraction forces
   c. failure to manipulate the electrode properly during welding
   d. insufficient heat applied to the weld
10. Insufficient penetration in a weld is usually caused by
    a. using the wrong type of electrode
    b. improper manipulation of the electrode
    c. too low a heat input
    d. not enough shielding over the crater

Short Answer

Write the correct word or words in the blank.

11. The load or force which tends to deform a material is called a _____.
12. The deformation which results because of an excessive load is known as a _____.
13. The point to which a material can be stretched and still return to its normal condition is referred to as _____.
14. The property which enables a material to resist forces that tend to pull it apart is called _____.
15. What is an example of a metal that has high compression strength but only moderate tensile strength? _____.
16. The property that enables a metal to withstand twisting loads is called _____.
17. What is the property of a material that enables it to absorb a large amount of energy without breaking? _____.
18. If a metal is able to withstand compression forces without being deformed, it is said to be _____.

UNIT 4
Shielded Metal-Arc Electrodes

There are many different kinds and sizes of electrodes and unless the correct one is selected you will have difficulty in doing a good welding job.

In general, all electrodes are classified into five main groups: *mild steel, high-carbon steel, special alloy steel, cast iron, and non-ferrous.* The greatest range of arc welding is done with electrodes in the *mild steel* group. Special alloy steel electrodes are made for welding various kinds of steel alloys. Cast iron electrodes are used for welding cast iron, and non-ferrous electrodes for welding such metals as aluminum, copper, and brass. In this unit we will discuss mild steel electrodes. Other types of electrodes will be covered in subsequent units dealing with the welding of special metals.

WHAT IS AN ELECTRODE?

An electrode is a coated metal wire having approximately the same composition as the metal to be welded. When the current is produced by the generator or transformer and flows through the circuit to the electrode, an arc is formed between the end of the electrode and the work. The arc melts the electrode and the base metal. The melted metal of the electrode flows into the molten crater and forms a bond between the two pieces of metal being joined.

Electrodes are not only manufactured to weld different metals; they are also designed for DC or AC welding machines. A few electrodes work equally well on either DC or AC. Then too, electrode usage depends on the welding position. Some electrodes are best suited for flat position welding, others are intended primarily for horizontal and flat welding, and some types are used for welding in any position.

Shielded electrodes have heavy coatings of various substances such as cellulose sodium, cellulose potassium, titania sodium, titania potassium, iron oxide, iron powder as well as several other ingredients. Each of the substances in the coating is intended to serve a particular function in the welding process. Thus it may:

1. Act as a cleansing and deoxidizing agent in the molten crater.
2. Release an inert gas to protect the molten

32 Arc Welding

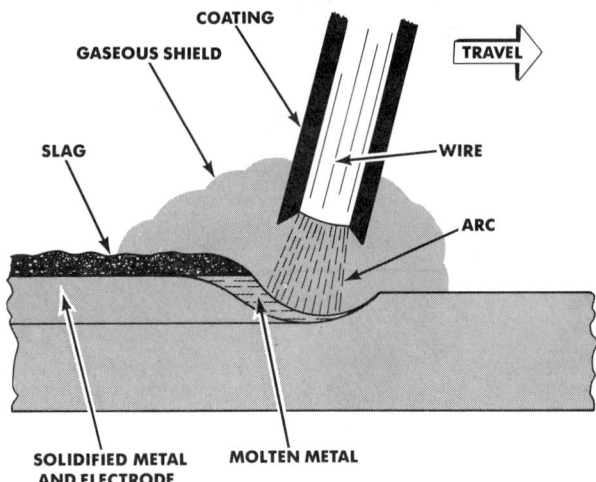

Fig. 4-1. Cross-section of a coated electrode in the process of welding. (The Lincoln Electric Co.)

metal from atmospheric oxides and nitrides. See Fig. 4-1. Since oxygen and nitrogen weaken a weld if allowed to come in contact with the molten metal, the exclusion of these contaminants is important.

3. Form a slag over the deposited metal which further protects the weld until the metal cools sufficiently so it is no longer affected by atmospheric contamination. The slag also slows the cooling rate of the deposited metal, thereby permitting a more ductile weld to form.

4. Provide easier arc starting, stabilize the arc better, and reduce splatter.

5. Permit better penetration and improve the X-ray quality of the weld.

The coating of some electrodes contains powered iron which converts to steel and becomes a part of the weld deposit. The powdered iron also helps to increase the speed of welding and to improve the weld appearance.

A group of electrodes known as low-hydrogen electrodes have coatings high in limestone and other ingredients with low-hydrogen content, such as calcium fluoride, calcium carbonate, magnesium-aluminum-silicate, and ferrous alloys. These electrodes are used to weld high-sulfur and high-carbon steels that have a great affinity for hydrogen which often causes porosity and underbead cracking in a weld.

Identifying Electrodes

You will find that electrodes are referred to by a manufacturer's trade name, or by an AWS identification symbol, such as E-6010, E-7010, E-8010, etc. The *prefix E* identifies the electrode for electric arc welding, as illustrated in Fig. 4-2. The *first two digits* in the symbol designate the minimum allowable tensile strength of the deposited weld metal in thousands of pounds per square inch (psi). For example, the 60 series electrodes have a minimum pull strength of 60,000 psi; the 70 series, a strength of 70,000 psi.

The *third digit* of the symbol indicates possible welding positions. Three numbers are used for this purpose: 1, 2 and 3. Number 1 is for an electrode which can be used for welding in any position. Number 2 represents an electrode restricted for welding in horizontal and flat positions only. Number 3 represents an electrode to be used in the flat position only.

The *fourth digit* of the symbol simply shows some special characteristic of the electrode, such as type of coating, weld quality, type of arc, and amount of penetration. The fourth digit may be 0, 1, 2, 3, 4, 5, 6, 7, or 8.

Thus for mild steel, the complete classification number E-6010 would signify an electrode that (a) has a minimum tensile strength of 60,000 for the as-welded deposited weld metal, (b) is usable in all welding positions, and (c) can be used with DC reverse polarity only. See Fig. 4-2.

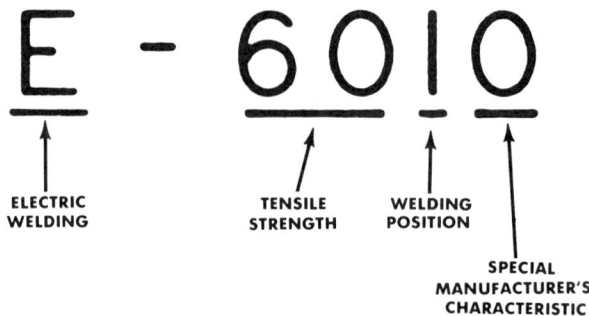

Fig. 4-2. Each of the letters and numbers used to classify electrodes has a specific meaning.

Selecting the Correct Electrode

The ideal electrode is one that will provide good arc stability, smooth weld bead, fast deposition, minimum spatter, maximum weld strength, and easy slag removal. To achieve these characteristics seven factors should be considered in selecting an electrode. These are:

Properties of the base metal. A top quality weld should be as strong as the parent metal. This means that the electrode to be used must produce a weld metal with approximately the same mechanical properties as the parent metal.

Electrodes are available for welding different classifications of metal. Thus some electrodes are designed to weld carbon steels, others are best suited for low-alloy steels and some are intended specifically for high-strength alloy steels. Therefore, in undertaking any welding operation, the first consideration is to check the type of metal to be welded and then select an electrode that is recommended for that metal. Most welding supply distributors are able to provide this information.

Electrode diameter. As a rule, an electrode is never used that has a diameter larger than the thickness of the metal to be welded. Some operators prefer larger electrodes because they permit faster travel along the joint and thus speed up the welding operation; but this requires considerable skill.

When making vertical or overhead welds, $3/16''$ is the largest diameter electrode that should be used regardless of plate thickness. Larger electrodes make it too difficult to control the deposited metal. Ordinarily, a *fast-freeze* type of electrode is best for vertical and overhead welding. See Table 4-1.

The diameter of the electrode is also influenced by the factors of the joint design. Thus, in a thick metal section with a narrow V, a small diameter electrode is always used to run the first weld bead or root pass. This is done to insure thorough penetration at the root of the weld.

TABLE 4-1. ELECTRODE CHARACTERISTICS.

TYPE	AWS CLASS	CURRENT TYPE	WELDING POSITION	WELD RESULTS
Mild steel	E-6010	DCR	F, V, OH, H	Fast freeze, deep penetrating, flat beads, all-purpose welding
	E-6011	DCR, AC	F, V, OH, H	
	E-6012	DCS, AC	F, V, OH, H	Fill-freeze, low penetration, for poor fit-up, good bead contour, minimum spatter
	E-6013	DCR, DCS, AC	F, V, OH, H	
	E-6020	DCR, DCS, AC	F, H	Fast-fill, high deposition, deep groove welds, single pass
Iron powder	E-6027	DCR, DCS, AC	F, H	Fast-fill, high deposition, deep penetration
	E-7014	DCR, DCS, AC	F, V, OH, H	Fill-freeze (combination of fast-fill and fast-freeze) low penetration, high speed
	E-7024	DCR, DCS, AC	F, H	Fast-fill, high deposition, single and multiple pass
Low hydrogen	E-7016	DCR, AC	F, V, OH, H	Welding of high-sulfur and high-carbon steels that tend to develop porosity and crack under weld bead
	E-7018	DCR, AC	F, V, OH, H	
	E-7028	DCR, AC	F, H	

DCR—Direct Current Reverse Polarity
DCS—Direct Current Straight Polarity
AC —Alternating Current
F —flat, V—vertical, OH—overhead, H—horizontal

Successive passes are then made with larger diameter electrodes.

Joint design and fit-up. Joints with insufficient beveled edges require deep penetrating, fast-freeze electrodes. Some electrodes have this particular digging characteristic and may require more skillful electrode manipulation by the operator. On the other hand, joints with open gaps need a mild penetrating fill-freeze electrode that rapidly bridges gaps. See Table 4-1 for variety of electrode characteristics.

Welding position. The position of the weld joint is an important factor in the type of electrode to be used. Some electrodes produce better results when the welding is done in a flat position. Other electrodes are designed for vertical, horizontal, and overhead welding.

Welding current. Electrodes are made for use with either AC current or DC current reverse polarity or DC current straight polarity, although some electrodes function as well on both AC and DC current.

Production efficiency. Deposition rate is extremely significant in any production work. The faster a weld can be made the lower the cost. Not all electrodes have a high-speed high-current rating that also produce smooth, even bead ripples. Unless electrodes are noted for a fast deposition rate, they may prove very difficult to handle when used at high speed travel.

Deposition Classification of Electrodes

Electrodes for welding mild steel are sometimes classified as fast-freeze, fill-freeze, and fast-fill[1] (see Table 4-1). The *fast-freeze* electrodes are those which produce a snappy, deep penetrating arc and fast-freezing deposits. They are commonly called reverse polarity electrodes even though some can be used on AC. These electrodes have little slag and produce flat beads. They are widely used for all types and all positions of welding for both fabrication and repair work.

Fill-freeze electrodes have a moderately forceful arc and deposit rate between those of the fast-freeze and fast-fill electrodes. They are commonly called straight-polarity electrodes even though they may be used on AC. These electrodes have complete slag coverage and weld beads with distinct, even ripples. They are the general-purpose electrodes for production shop and are also widely used for repair work. They can be used in all positions, though the fast-freeze electrodes are preferred for vertical and overhead welding.

The *fast-fill* group includes the heavy coated, iron powder electrodes with soft arc and fast deposit rate. These electrodes have a heavy slag and produce exceptionally smooth weld beads. They are generally used for production welding where all work can be positioned for downhand (flat) welding.

Conserving and Storing Electrodes

Most electrodes are costly; therefore, every bit of the electrode should be consumed. Do not discard stub ends until they are down to only 1½" to 2" long. See Fig. 4-3.

Always store electrodes in a dry place at a normal room temperature and 50 percent maximum relative humidity. When exposed to moisture, the coating has a tendency to disintegrate. In storing electrodes, be sure they are not bumped, bent, or stepped on, since this will remove the coating and render the electrode useless.

Fig. 4-3. Do not discard electrodes of this length.

---

1. The Lincoln Electric Company.

Types of Mild Steel Electrodes

E-6010. This is an all-position, fast-freeze electrode. It is suitable only on DC machines with reversed polarity, and is designed primarily for welding mild and low-alloy steels. It should be used only where there is an absolutely good fit-up. The E-6010 electrode has wide applications in ship construction, buildings, bridges, tanks, and piping. Table 4-2 shows the amperage settings for different sizes of this electrode.

E-6011. The E-6011 electrode is similar to the E-6010, except that it is made especially for AC machines. Although the electrode can be used on DC machines with reversed polarity, it does not work quite as well as the E-6010. The amperage setting used with it is slightly lower than that used with E-6010. See Table 4-3.

E-6012. This is a fill-freeze electrode and may be used on either DC or AC welders. When employed on DC welders the current must be set for straight polarity. The electrode provides medium penetration, a quiet type arc, slight spatter, and dense slag. Although it is considered an all-position electrode, it is used in greater quantities for flat and horizontal position welds. This electrode is especially useful to bridge gaps under conditions of poor *fit-up work,* that is, joints where the edges do not fit closely together. Higher currents can be used with the E-6012 electrodes than with any other type of all-position electrodes. See Table 4-4.

TABLE 4-2. CURRENT SETTINGS FOR E-6010 ELECTRODES.

ELECTRODES DIA (INCHES)	AMPERES*
3/32	60–90
1/8	80–120
5/32	110–160
3/16	150–200
7/32	175–250
1/4	225–300
5/16	250–450

*These ranges may vary slightly for electrodes made by different manufacturers

TABLE 4-3. CURRENT SETTINGS FOR E-6011 ELECTRODES.

ELECTRODES DIA (INCHES)	AMPERES*
3/32	50–90
1/8	80–130
5/32	120–180
3/16	140–220
7/32	170–250
1/4	225–325

*These ranges may vary slightly for electrodes made by different manufacturers

TABLE 4-4. CURRENT SETTINGS FOR E-6012 ELECTRODES.

ELECTRODE DIA (INCHES)	AMPERES*
3/32	40–90
1/8	80–120
5/32	120–190
3/16	140–240
7/32	180–315
1/4	225–350

*These ranges may vary slightly for electrodes made by different manufacturers

E-6013. Electrodes of this type are very similar to E-6012 with a few slight exceptions. Slag removal is better and the arc can be maintained easier, especially with small diameter electrodes. This permits better operation with lower open-circuit voltage. The bead deposited is noticeably flatter and smoother but with shallower penetration than that of the E-6012 class. Although the electrode is used particularly for welding sheet metal, it has many other applications. It works well in all positions and it functions very well on AC welders. When used with DC machines the polarity may be straight or reverse. Current settings for E-6013 electrodes are shown in Table 4-5.

TABLE 4-5. CURRENT SETTINGS FOR E-6013 ELECTRODES.

ELECTRODE DIA (INCHES)	AMPERES*
1/16	20–40
5/64	25–50
3/32	30–80
1/8	80–120
5/32	120–190
3/16	140–240
7/32	225–300
1/4	250–350

*These ranges may vary slightly for electrodes made by different manufacturers

TABLE 4-6. CURRENT SETTINGS FOR E-7014 ELECTRODES.

ELECTRODE DIA (INCHES)	AMPERES*
3/32	80–110
1/8	110–150
5/32	140–190
3/16	180–260
7/32	250–325
1/4	300–400
5/16	400–500

*These ranges may vary slightly for electrodes made by different manufacturers

TABLE 4-7. CURRENT SETTINGS FOR E-7024 ELECTRODES.

ELECTRODE DIA (INCHES)	AMPERES*
3/32	90–120
1/8	120–150
5/32	180–230
3/16	250–300
7/32	300–350
1/4	350–400
5/16	400–500

*These ranges may vary slightly for electrodes made by different manufacturers

Iron Powder Electrodes

Iron powder electrodes are those which contain a high content of iron powder. They are designed for welding mild steels where high speed and fast deposition rate are required. The two principal types are E-7014, and E-7024. All of them produce low spatter with easy slag removal. Typical application includes railroad cars, earth-moving equipment, positioned welds in pressure vessels, piping, and ships. The E-7014 and E-7024 are often used where higher strength joints are necessary.

E-7014. This is a fast-fill and fast-freeze electrode where high speed is necessary. It may be used in all positions with AC or DCR and DCS current. The E-7014 electrode deposits much more metal than an E-6012 or E-6013 type. It is particularly effective in vertical downhill welding. See Table 4-6 for approximate current settings.

E-7024. The E-7024 is a fast-fill electrode which provides exceptional economy for single or multiple pass welds and is excellent on build-up applications because of the high deposition rate and easy slag removal. It is recommended only for flat and horizontal positions with AC or DC straight or reverse polarity. See Table 4-7 for approximate current settings.

Low-Hydrogen Electrodes

Low-hydrogen electrodes are designed for welding high-sulfur and high-carbon steels. When such steels are welded they tend to develop porosity and cracks under the weld bead because of the hydrogen absorption from arc atmospheres. Low-hydrogen electrodes were developed to prevent the introduction of hydrogen in the weld.

The basic low-hydrogen electrodes are E-7018, and E-7028.

E-7018. The E-7018 is a low-hydrogen type

TABLE 4-8. CURRENT SETTINGS FOR E-7018 ELECTRODES.

ELECTRODE DIA (INCHES)	AMPERES*
3/32	70–120
1/8	100–150
5/32	120–200
3/16	200–275
7/32	275–350
1/4	300–400

*These ranges may vary slightly for electrodes made by different manufacturers

TABLE 4-9. CURRENT SETTINGS FOR E-7028 ELECTRODES.

ELECTRODE DIA (INCHES)	AMPERES*
5/32	175–250
3/16	250–325
7/32	300–400
1/4	375–475

*These ranges may vary slightly for electrodes made by different manufacturers

electrode that also contains iron powder. It is a high-speed fast-deposition rate electrode designed to pass the most severe X-ray requirements when applied in all welding positions, using either AC or DC reverse polarity current. Its puddle fluidity permits gases to escape when the lowest currents are used for out-of-position welding. See Table 4-8 for approximate current settings for low-hydrogen electrodes.

E-7028. The E-7028 is a low-hydrogen electrode with a heavy iron powder type covering and is considered the counterpart of E-7018 but for flat and horizontal positions only. See Table 4-9 for approximate current settings.

SELF QUIZ

Correct answers are listed in the back of the book.

True and False

Circle the letter T if the statement is true or the letter F if the statement is false.

1. T F The coating on an electrode produces a gas which protects the weld from atmospheric oxides and nitrides.
2. T F Oxygen and nitrogen, when allowed to come in contact with molten metal, will weaken a weld.
3. T F The coating that forms over a deposited weld bead is called slag.
4. T F The purpose of the slag over a weld is to reduce the cooling rate of the deposited metal.
5. T F The type of coating used on an electrode does not affect its starting qualities.
6. T F Some electrode coatings contain iron powder, which converts to steel and becomes part of the weld deposit.
7. T F Iron powder type electrodes do not have a high deposition rate.
8. T F Electrodes classified as E-60xx have higher tensile strength than those classified as E-70xx.

9. T F In the AWS classification, the third digit identifies the manufacturer of the electrode.
10. T F In most welding situations, the diameter of the electrode should never exceed the thickness of the metal to be welded.
11. T F Some experienced welders often use larger diameter electrodes to get greater weld strength.
12. T F Fast-freeze type electrodes have deep penetrating qualities.
13. T F Fill-freeze electrodes are usually intended for welding joints with poor fit-up.
14. T F Fast-fill electrodes are best for fast downhand welding where only one pass is required.
15. T F Low-hydrogen type electrodes are designed for welding common low carbon steels.
16. T F Some electrodes may be used with either DC or AC current.
17. T F Each type of electrode is designed to be used for one welding position.

Identification

Write the correct answers in the blanks provided.

18. Identify the following abbreviations.

DCR _____

DCS _____

AC _____

F _____

V _____

OH _____

H _____

19. Identify the following electrode markings.

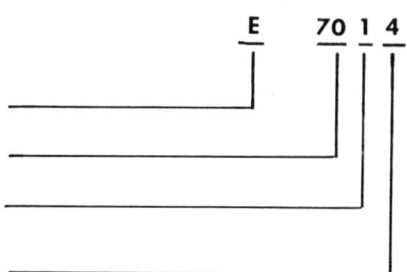

20. Give the electrode series and the main characteristics (position, purpose, etc.) for the following electrode groups:

A. Fast-fill

B. Fast-freeze

C. Fill-freeze

D. Low-hydrogen

Electrode Series

A. _____

B. _____

C. _____

D. _____

Characteristics

A. _____

B. _____

C. _____

D. _____

UNIT 5
Shielded Metal-Arc Welding Equipment

Shielded metal-arc welding, sometimes referred to as metallic-arc welding, or just stick welding, is widely used in the construction of many products, ranging from steamships, tanks, locomotives, and automobiles to small household appliances. Arc welding machines today are designed to join light and heavy gage metals of all kinds. The process of arc welding not only simplifies the maintenance and manufacture of goods and machines, but it permits the skilled operator to perform welding operations quickly and easily.

Welding Current

The heat generated for welding comes from an arc which develops when electricity jumps across an air gap between the end of an electrode and the base metal. The air gap produces a high resistance to the flow of current and this resistance generates an intense arc heat which may be anywhere from 6000°F to 10,000°F (approximately 3300°C to 5500°C).

Welding current is provided by an AC or DC machine. The primary current (input) to a welding machine is either 220 or 440 volts. *Since voltage of this magnitude is always dangerous, extreme care must be taken to insure that the motor and frame are well grounded.*

The actual voltage used to provide welding current is low (18 to 36V) whereas high amperage is necessary to produce the heat required for welding. However, low voltage and high amperage for welding are not particularly dangerous if there is adequate grounding and proper insulation. Although they need not be feared, both should be treated with care to avoid any electrical accident.

Electrical Terms

To understand the correct operation of an electric arc welding machine, you must know a few basic electrical terms and electric principles. The following are especially important:

Alternating current (AC). An electrical current having alternating positive and negative values. In the first half cycle the current flows in one direction then reverses direction and for the next half cycle flows the opposite way. The rate of change is referred to as frequency. This frequency is indicated as 25, 40, 50 and 60 cycles per second. In the United States, alternating current is usually established at 60 cycles per second.

Direct current (DC). Electrical current which flows in one direction only.

Conductor. A conductor is any material in the form of a wire, cable or bus bar which allows a free passage of an electrical current.

Electrical circuit. Path taken by an electric current in flowing through a conductor from one terminal of the source of supply to the other. It starts from the negative terminal of the power supply where the current is produced, moves along the wire or cable to the load or working source, and then returns to the positive terminal. See Fig. 5-1.

Fig. 5-2. An ammeter shows the amount of current that is flowing. (Weston Electrical Instrument Corp.)

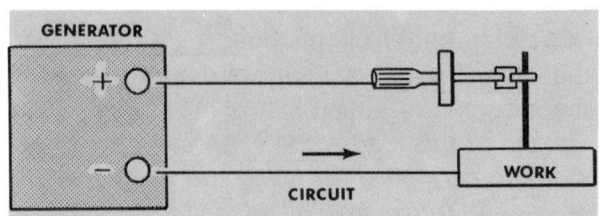

Fig. 5-1. An example of a simple electrical circuit.

Amperage. Amperes (abbreviated amp), or amperage, refers to the amount or rate of current that flows in a circuit. The instrument that measures this rate is called an ammeter, shown in Fig. 5-2.

Voltage. The force (emf, or electro-motive force) that causes current to flow in a circuit is known as voltage. This force is similar to the pressure used to make water flow in pipes. In a water system, the pump provides the pressure, whereas in an electrical circuit the power supply produces the force that pushes the current through the wires. Voltage does not flow; only

Fig. 5-3. A voltmeter measures the force of electricity flowing in a circuit. (Weston Electrical Instrument Corp.)

current flows. The force is measured in volts and the instrument used to measure voltage is called a voltmeter. See Fig. 5-3.

Shielded Metal-Arc Welding 41

Resistance. Resistance is the opposition of the material in a conductor to the passage of an electric current causing electrical energy to be transformed into heat.

Static electricity. Static electricity refers to electricity at rest or electricity that is not moving.

Dynamic electricity. Dynamic electricity is electricity in motion in an electrical current.

Voltage drop. Just as the pressure in a water system drops as the distance increases from the water pump, so does the voltage decrease as the distance increases from the generator. This fact is important to remember in using a welding machine, because if the cables are too long, there will be too great a voltage drop. When there is too much drop, the welding machine cannot supply enough current for welding.

Open-circuit voltage and arc voltage. Open-circuit voltage is the voltage produced when the machine is running and no welding is being done. This voltage varies from 50 to 100V. After the arc is struck, the voltage drops to what is known as the *arc or working voltage,* which is between 18 and 36V. An adjustment is provided to vary the open circuit voltage so welding can be done in different positions. See Figs. 5-4 and 5-5.

Polarity. Polarity indicates the direction of the current in a circuit. Since the current moves in one direction only in *DC welders,* polarity is important because for some welding operations the flow of current must be changed. When the electrode holder is fastened to the negative pole of the generator and the work to the positive pole, the polarity is negative, or more commonly referred to as *straight polarity.* See Fig. 5-6. If the

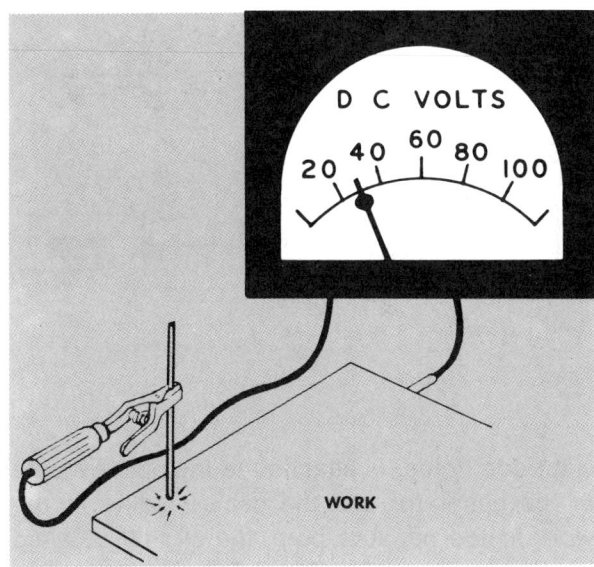

Fig. 5-5. Arc voltage is the voltage used when welding is in process.

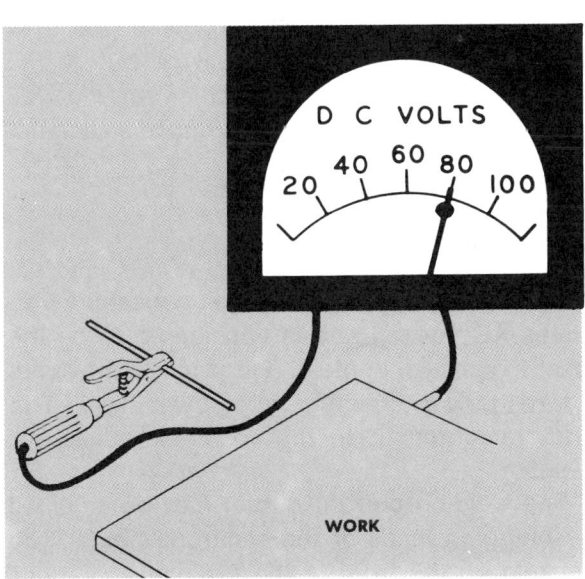

Fig. 5-4. When the machine is running and no welding is being done, you have an open circuit voltage.

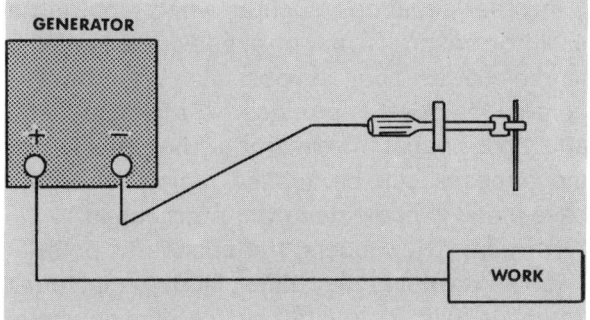

Fig. 5-6. This is how the circuit is arranged for straight polarity.

42 Arc Welding

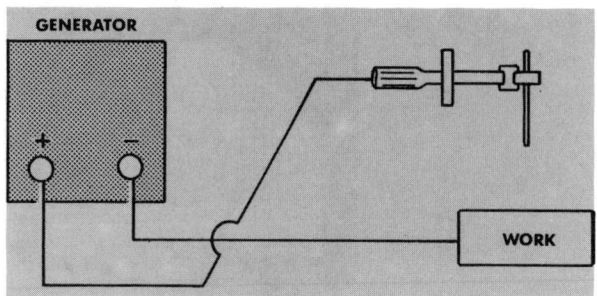

Fig. 5-7. This is how the cables are connected for reversed polarity.

electrode holder is attached to the positive pole of the generator and the cable leading to the work to the negative pole, the circuit is called *reversed polarity*. See Fig. 5-7.

Polarity has a direct relationship to the location of the liberated heat, since it is possible to control the amount of heat going into the base metal. By changing polarity the greatest heat can be concentrated where it is most needed.

For some types of welding situations, it is preferable to have more heat at the workpiece, because the area of the work is greater and more heat is required to melt the metal than the electrode. Thus, if large, heavy deposits are to be made, the work should be hotter than the electrode. For this purpose, straight polarity would be more effective.

On the other hand, in overhead welding it is necessary to quickly freeze the filler metal to help hold the molten metal in position against the force of gravity. By using reverse polarity, less heat is generated at the workpiece, thereby giving the filler metal greater holding power for out-of-position welding.

In other situations, such as when repairing a cast iron casting, it may be expedient to keep the workpiece as cool as possible. With reverse polarity less heat is produced in the base metal and more heat at the electrode. The result is that the deposits can be applied rapidly while the base metal is prevented from overheating.

On early DC welders, the change of polarity involved reversing the cables. Modern machines equipped with a *polarity switch* eliminate disconnecting the cables. Moving the switch to straight or reverse, changes the polarity.

Inasmuch as the current is constantly reversing in AC welders, polarity is of no consequence.

WELDING MACHINES

To supply the current for welding, three types of units are available: transformers, motor generators, and rectifiers. The source of power to run these welding machines may come from regular electrical lines, or gasoline and diesel engines. The gasoline or diesel run types are especially useful for field work where electrical power is not available.

Sizes of machines. Sizes of welding machines are rated according to their approximate amperage capacity at *60 percent duty cycle*, such as 150, 200, 250, 300, 400, 500 or 600. This amperage is the rated current output at the working terminal. *Thus a machine rated at 150 amperes can be adjusted to produce a range of power up to 150 amperes.*

Transformers

The transformer type of welding machine produces AC current and is considered to be the least expensive, lightest, and smallest machine. It takes power directly from a power supply line and transforms it to the voltage required for welding.

The welding current output may be adjusted by plugging leads of the electrode holder into sockets on the front of the machine in various locations, or by rotating a handwheel or crank. See Figs. 5-8 and 5-9.

Shielded Metal-Arc Welding   43

DC Welding Machines

Motor generators are designed to produce DC current in either straight or reverse polarity. The polarity selected for welding depends on the kind of electrode used and the material to be welded. A switch on the machine can be turned for straight or reverse polarity.

Present day motor generators for manual stick welding are usually of the constant current, dual control type.

With a dual control machine welding current is adjusted by two controls. One control provides an approximate or coarse setting. The second control is usually a rheostat that can be turned to provide a fine adjustment of the welding current and increase or decrease the heat. See Fig. 5-10.

On dual control machines the slope of the output current can be varied to produce a soft or

Fig. 5-8. Welding current output on some AC machines is regulated by plugging leads into sockets. (Miller Electric Manufacturing Co.)

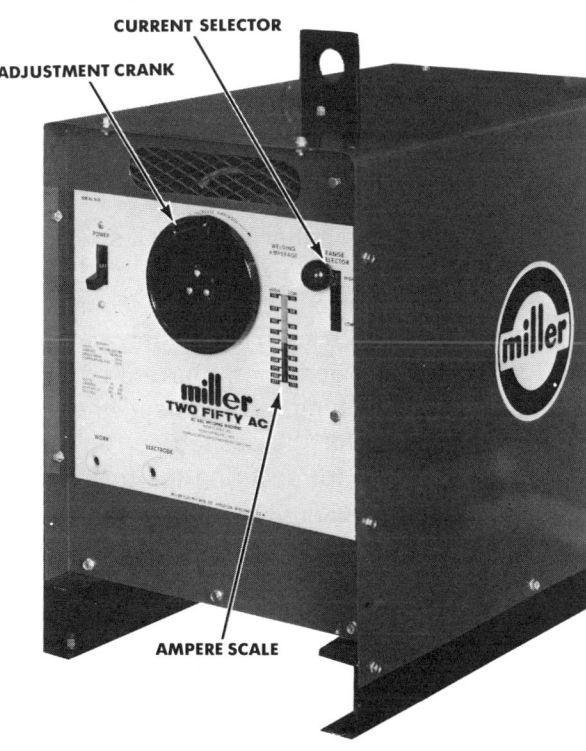

Fig. 5-9. An AC arc welder. Coarse adjustments are made by moving the electrode selector from the high to the low range tap. Fine adjustments are made by rotating the crank. (Miller Electrical Manufacturing Co.)

Fig. 5-10. Constant-current DC welding machine. (The Lincoln Electric Co.)

44 Arc Welding

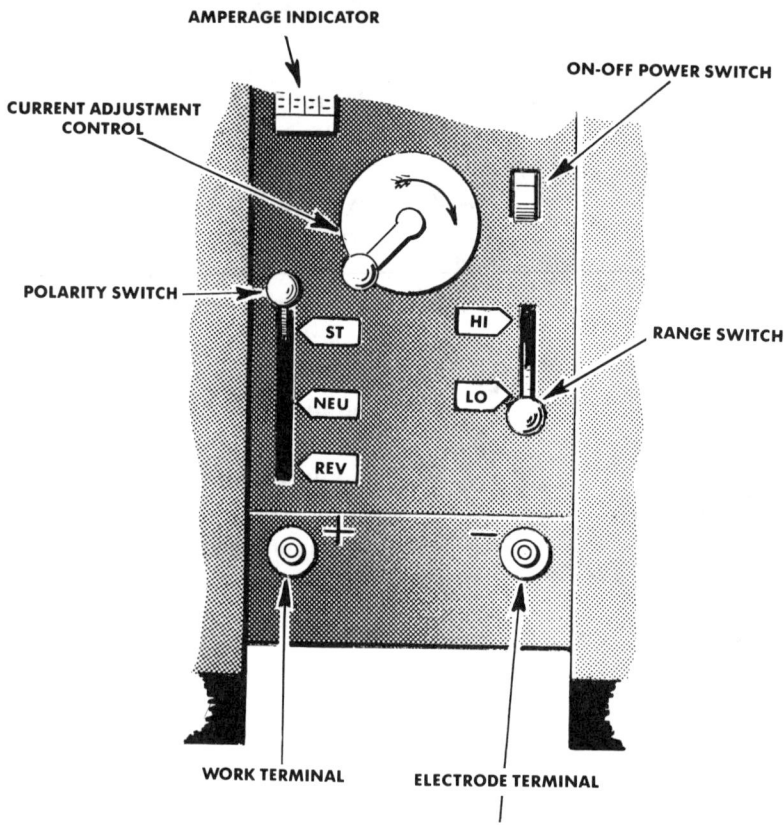

Fig. 5-11. DC/AC rectifier type adaptable to a variety of welding applications. (Airco)

harsh arc. By flattening the volt-amp curve (increasing the amperage) a digging arc can be obtained for deeper penetration. With a steeper curve (reduced amperage in relation to voltage) a soft or quiet arc results which is useful for welding light gage materials. In other words, a machine with dual control allows the greatest flexibility for welding materials of different thicknesses.

Rectifiers

Rectifiers are essentially transformers containing an electrical device which changes alternating current into direct current.

They are available to produce both DC and AC current. By turning a switch the output terminals can be changed to the transformer or to the rectifier to produce either AC or DC, straight or reverse polarity current.

The transformer rectifier is usually considered more efficient electrically than the generator and provides quiet operation. Current control is achieved by a switching arrangement where one switch can be set for the desired current range and a second dial for securing the fine adjustment desired for welding. See Fig. 5-11.

WELDING ACCESSORIES

Helmet and hand shield. An electric arc not only produces a brilliant light, but it also gives off invisible ultraviolet and infrared rays which are extremely dangerous to the eyes and skin.

CAUTION: You should never look at an arc with the naked eye!

Shielded Metal-Arc Welding

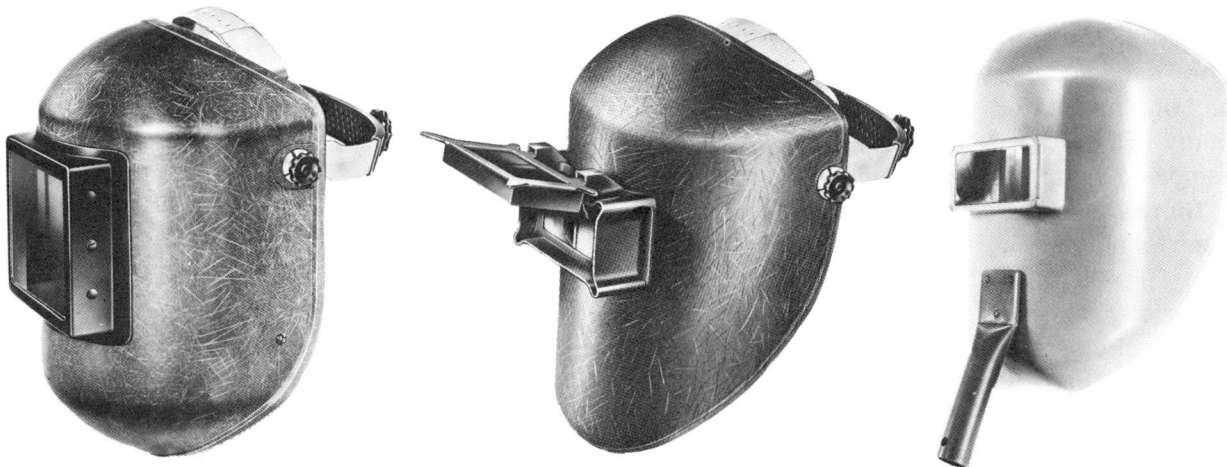

Fig. 5-12. Types of head shields used for arc welding.

To protect yourself from harmful rays, you must use either a helmet or hand shield as illustrated in Fig. 5-12.

The *helmet* fits over the head and can be swung upward or the lenses opened when you are not welding. The chief advantage of the helmet is that it leaves both hands free, thereby making it possible to hold the work and to weld at the same time.

The *hand shield* provides the same protection as the helmet, except that it is held in position by a handle. This shield is frequently used by an observer or a person who welds for only a short period. Both types of shields are equipped with special colored lenses that reduce the brilliancy of the light and screen out the infrared and ultraviolet rays. Lenses come in different shades, and the type used depends on the kind of welding done. In general, the recommended practice is as follows:

Shade 2 for resistance welding and stray light.

Shade 5 for light oxygen cutting and gas welding.

Shade 6 and 7 for arc welding up to 30 amperes, oxygen cutting and medium gas welding.

Shade 8 for arc welding beyond 30 and up to 75 amperes and heavy gas welding.

Shade 10 for arc cutting and welding beyond 75 and up to 200 amperes.

Shade 12 for arc welding and cutting beyond 200 and up to 400 amperes.

Shade 14 for arc welding and cutting over 400 amperes.

During the welding process, small particles of metal fly upward from the work and may lodge on the lens. Colored lenses are protected by use of *clear glass* or *plastic cover plates.*

These clear glasses are inexpensive and can be purchased from any welding supply dealer. *Since you must have clear vision at all times during welding, always replace the cover glass when enough spatter has accumulated on it to interfere with your vision.*

Goggles. Goggles, as shown in Fig. 5-13, are worn when chipping slag from a weld. In arc welding a thin crust forms on the deposited

Fig. 5-13. Goggles must always be used when removing slag from a weld.

bead. This substance, known as slag, must be removed. While removing slag, tiny particles are often deflected upward. Without proper eye protection, these particles may cause a serious eye injury. Therefore, *always wear goggles when removing slag from a weld.*

Gloves. Another important item you will need for arc welding is a pair of gloves. *You must wear gloves to protect your hands from the ultraviolet rays and spattering hot metal.*

Several different kinds of gloves are available. As a rule, the leather gauntlet gloves shown in Fig. 5-14 provide ample protection. Regardless of the type used, they should be flexible enough to permit proper hand movement, yet not so thin as to allow the heat to penetrate easily.

Fig. 5-15. An apron provides protection for your clothes. (The Lincoln Electric Co.)

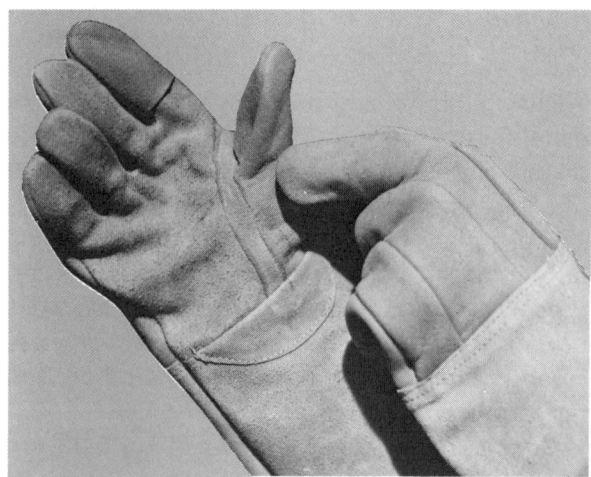

Fig. 5-14. Always wear gloves when arc welding. (The Lincoln Electric Co.)

Apron. It is a good idea to wear an apron when learning to arc weld. Otherwise, spattering metal might ruin your clothes. Since the spattering particles are hot, a leather apron offers the best protection. A typical welding apron is illustrated in Fig. 5-15.

Most experienced welders seldom wear an apron on the job except in situations where there may be an excessive amount of metal spatter resulting from awkward welding positions. Usually they wear suitable coveralls (fire retardant) to protect their clothing.

Coveralls or ordinary clothing should be sufficiently heavy to prevent infrared and ultraviolet rays from penetrating the skin. Cuffs on overalls should be turned down or eliminated, and pockets removed so they will not serve as lodging places for falling globules of molten metal. Sleeves and collars should be kept buttoned. Ankle-type shoes are preferred over the Oxford type.

Electrode holder. To do a good welding job, a properly designed electrode holder is essential. The holder is a handle-like tool attached to the cable that holds the electrode during welding. See Fig. 5-16.

A well designed holder can be identified by these features:

1. It is reasonably light, to reduce excessive fatigue while welding.
2. It does not heat too rapidly.
3. It is well balanced.
4. It receives and ejects the electrodes easily.
5. All exposed surfaces, including the jaws, are protected by insulation.

The jaws of some holders are not insulated. *When using a holder with uninsulated jaws, never lay it on the bench plate while the machine is running because it will cause a flash.*

Cleaning tools. To produce a strong welded joint, the surface of the metal must be free of all

Shielded Metal-Arc Welding   47

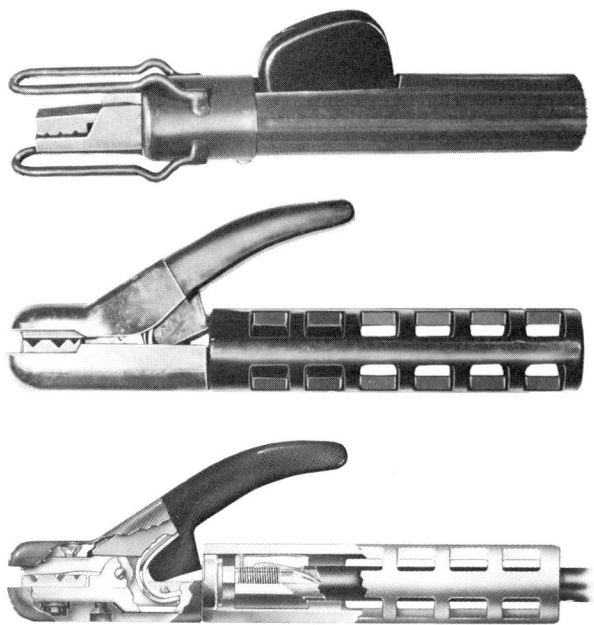

Fig. 5-16. Insulated and ventilated handles on rod holders. (The Lincoln Electric Co.)

foreign matter, such as rust, oil, and paint. A steel brush is used for cleaning purposes.

After a bead is deposited on the metal, the slag which covers the weld is removed with a chipping hammer, which is pictured in Fig. 5-17. The chipping operation is followed by additional wire brushing. Complete removal of slag is especially important when several passes must be made over a joint. Otherwise, gas holes will form in the bead, resulting in porosity which weakens the weld.

Cables. The cables carry the current to and from the work. One cable runs from the welding machine to the holder and the other cable is attached to the work or bench. The cable connected to the work is called the *ground cable.* Thus when the welding machine is turned on and the electrode in the holder comes in contact with the work, a circuit is formed, allowing the electricity to flow.

It is important to use the correct diameter cable specified for the welding machine. If the cable is too small for the current, it overheats and a lot of power is lost. Furthermore, a larger cable is necessary to carry a required voltage any distance from the machine. Otherwise, there will be an excessive voltage drop. Even with larger diameter cables, you must take precaution not to exceed the recommended lengths, because the voltage drop will lower the efficiency of your welding.

All cable connections should be tight, because any loose connection will cause the cable or

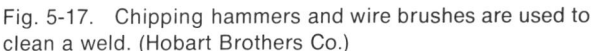

Fig. 5-17. Chipping hammers and wire brushes are used to clean a weld. (Hobart Brothers Co.)

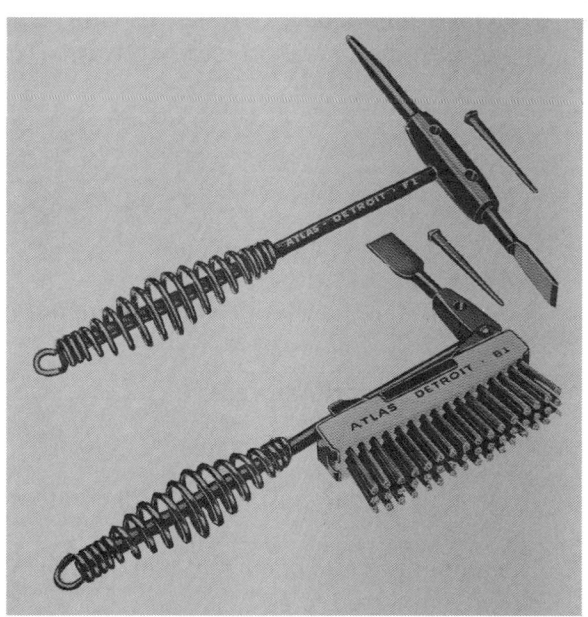

48 Arc Welding

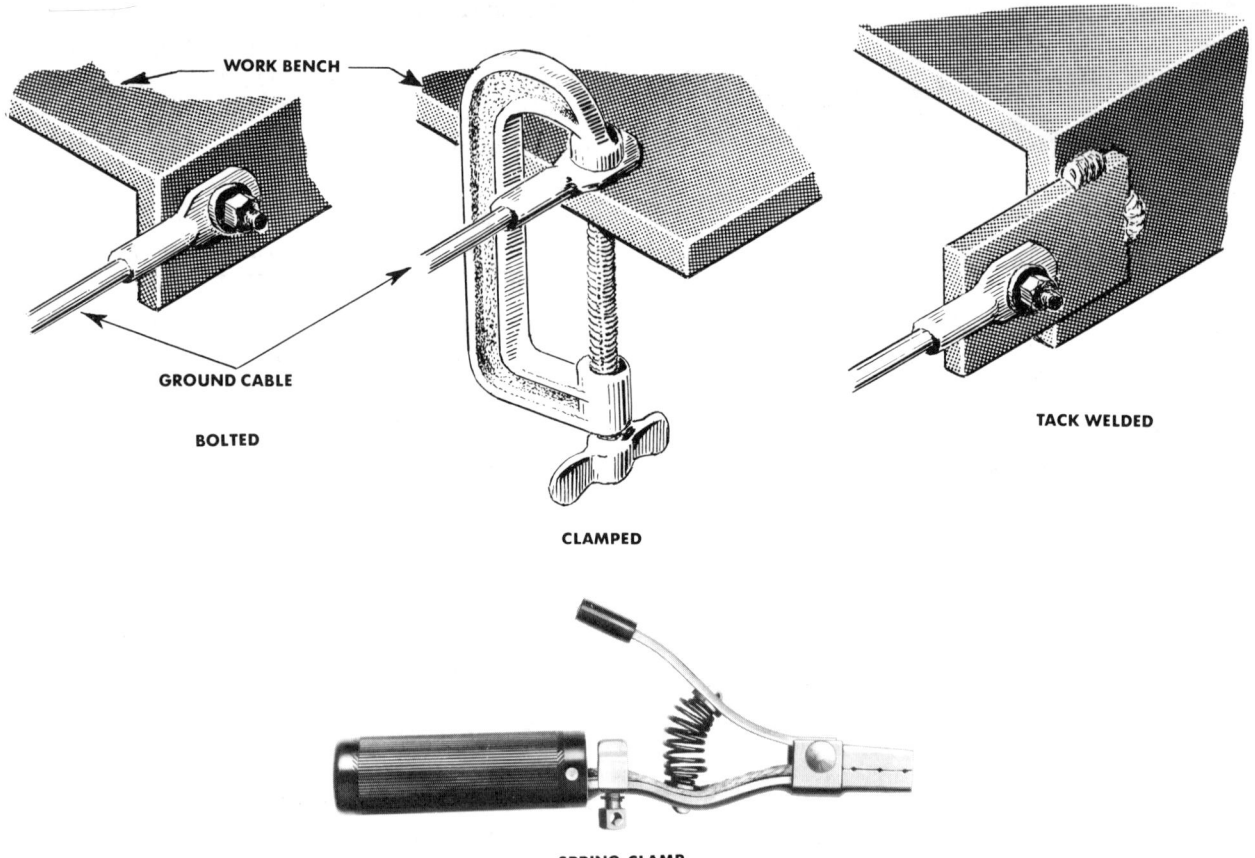

Fig. 5-18. Several connections can be used to provide a proper ground.

clamp to overheat. A loose connection may even produce arcing at the connection.

Ground connections. Proper ground connections can be made in several ways. The ground cable can be fastened to the work or bench by a *C* clamp, a special ground clamp, or by bolting or welding the lug on the end of the cable to the bench. See Fig. 5-18.

SELF QUIZ

Correct answers are listed in the back of the book.

Short Answers

Write the answers on the blanks.

1. What kind of current do motor generators produce? _____
2. The amount or rate of current that flows in a circuit is referred to as _____.
3. The force that causes current to flow is called _____.
4. When a machine is running and no welding is being done, the voltage is known as _____.

5. The normal range of the arc or working voltage is _____.
6. The term which indicates the direction of current flow is _____.
7. When the cable leading to the work is attached to the negative pole, the current flows from the work to the electrode, and the polarity is said to be _____.
8. When the cable leading to the work is attached to the positive pole, the current flows from the electrode to the work, and the polarity is said to be _____.
9. Does straight or reverse polarity produce the greatest amount of heat at the workpiece? _____
10. With what type of welding machine is polarity of no consequence? _____
11. Sizes of welding machines are rated according to their amperage capacity. Is this true or false? _____
12. Rectifier welding machines are available to produce both DC and AC current. Is this true or false? _____
13. Certain types of rays produced by an electric arc are harmful to the eyes. Which are those rays? _____
14. What governs the shade of lenses to be used while welding? _____
15. For safety reasons, what must you wear when removing slag from a weld? _____
16. What will happen if a holder with uninsulated jaws is placed on the bench while the machine is running? _____

Special Assignment

Before you start any welding assignment check your equipment to see that everything is right for welding. Use the Check List below for this purpose. If you find anything wrong, check with your instructor.

Welding Equipment Check List

Place a check mark (✓) after each item.

1. The machine is properly grounded. _____
2. The cable ground clamp is properly secured. _____
3. The main power switch is turned on for welding. _____
4. The machine, if a motor generator, is set for the correct polarity. _____
5. The amperage control is set at the approximate current for the electrode to be used. _____
6. The electrode holder is in good condition. _____
7. The bench top is clean and dry. _____
8. Welding gloves are available for use. _____
9. I am wearing proper protective clothing for welding. _____
10. The clear cover glasses over the helmet lens are relatively free of metal spatter. _____
11. The welding area where I am to work is properly shielded and there is ample ventilation. _____
12. Slag removal equipment is available. _____

UNIT 6
Principles of Shielded Metal-Arc Welding

Learning to arc weld involves mastery of a specific series of operations. Skill in performing these operations requires practice. Once this skill has been acquired, the operations can be applied on any welding job. The first basic operation is learning to strike an arc and running straight beads.

Gripping the Electrode Holder

Place the bare end of the electrode in the holder as shown in Fig. 6-1. By gripping the electrode near the end, most of the coated portion can be used. *Always keep the jaws of the holder clean* to insure good electrical contact with the electrode. *Be careful not to touch the welding bench with an uninsulated holder, as this will cause a flash. When not in use, hang the holder in the place provided for it.*

Grip the holder lightly in your hand. If you hold it too tightly, your hand and arm will tire quickly. Whenever possible, drape the cable over the shoulder or knee to lessen its drag on the holder.

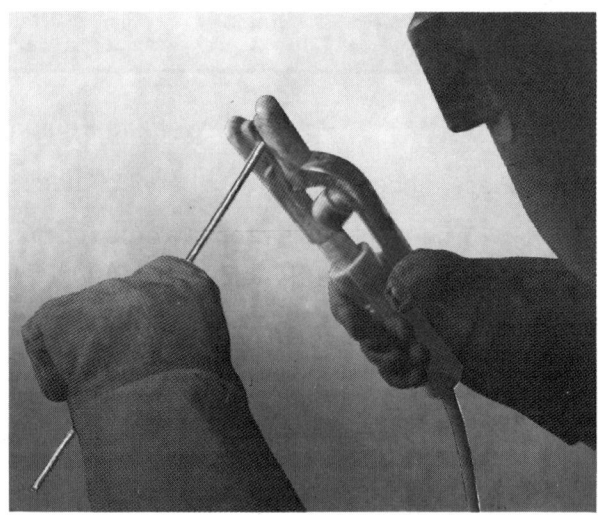

Fig. 6-1. How to grip the electrode. (Hobart Brothers Co.)

Striking the Arc

There are two methods which can be used to start, or strike the arc—the *tapping* and the *scratching motion.* The tapping method is the one preferred by experienced welders, whereas the scratching motion is found to be easier for the beginner.

In the tapping motion, the electrode is brought straight down and withdrawn instantly, as shown in top of Fig. 6-2. With the scratching method, the electrode is moved at an angle to the plate in a scratching motion much as in striking a match shown in bottom of Fig. 6-2. Regardless of which motion you use, upon contact with the plate, promptly raise the electrode a distance equal to the diameter of the electrode. Otherwise, the electrode will stick to the metal. If it is allowed to remain in this position with the current flowing, the electrode will become red hot. *Should the electrode weld fast to the plate, break it loose by quickly twisting or bending the holder. If it should fail to dislodge, disengage the electrode by releasing it from the holder.*

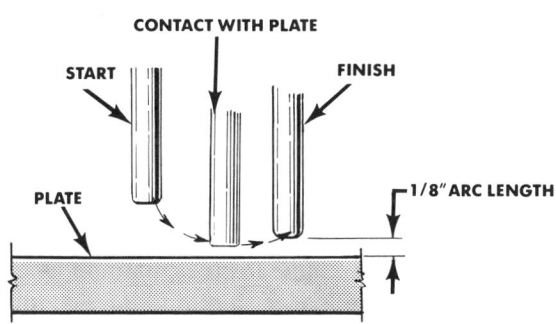

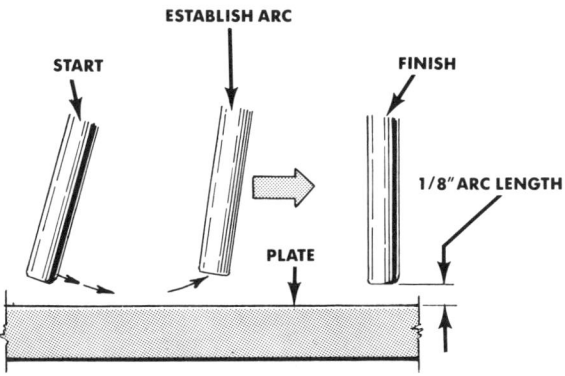

Fig. 6-2. There are two methods of starting, or striking, a welding arc.

Depositing Weld Beads

To secure a weld that has proper penetration, you must keep in mind the following five factors: (1) Correct electrode, (2) Correct arc length, (3) Correct current, (4) Correct travel speed, and (5) Correct electrode angle.

Correct electrode. The choice of an electrode involves such items as position of the weld, properties of the base metal, diameter of the electrode, type of joint, and current value. Since many different kinds of electrodes are manufactured, you must know the results that can be expected from different electrodes. If the characteristics of the electrodes are known, then you have greater assurance that a correct weld will be made. Without the right kind of electrode it is almost impossible to get the results desired, regardless of the welding technique used.

Correct arc length. If the arc is too long, the metal melts off the electrode in large globules which wobble from side to side as the arc wavers. This produces a wide, spattered, and irregular bead without sufficient fusion between the base metal and the deposited metal. An arc that is too short fails to generate enough heat to melt the base metal properly. Furthermore, the electrode sticks frequently, producing a high, uneven bead with irregular ripples.

The length of the arc depends on the size of electrode used and the kind of welding done. Thus for small diameter electrodes, a shorter arc is necessary than for larger electrodes. As a rule, the length of the arc should be approximately equal to the diameter of the electrode. For example, an electrode 1/8" in diameter should have an arc length of about 1/8". You will find, too, that a shorter arc is required for vertical and overhead welding than for most flat position welding, because it gives better control of the puddle.

The use of a short arc is important because it prevents impurities from entering a weld. A long arc allows the atmosphere to flow into the arc stream, thereby permitting impurities of nitrides and oxides to form. Moreover, when the arc is too long, heat from the arc stream is dissipated too rapidly, causing considerable metal spatter.

Correct current. If the current is too high, the electrode melts too fast and the molten pool is

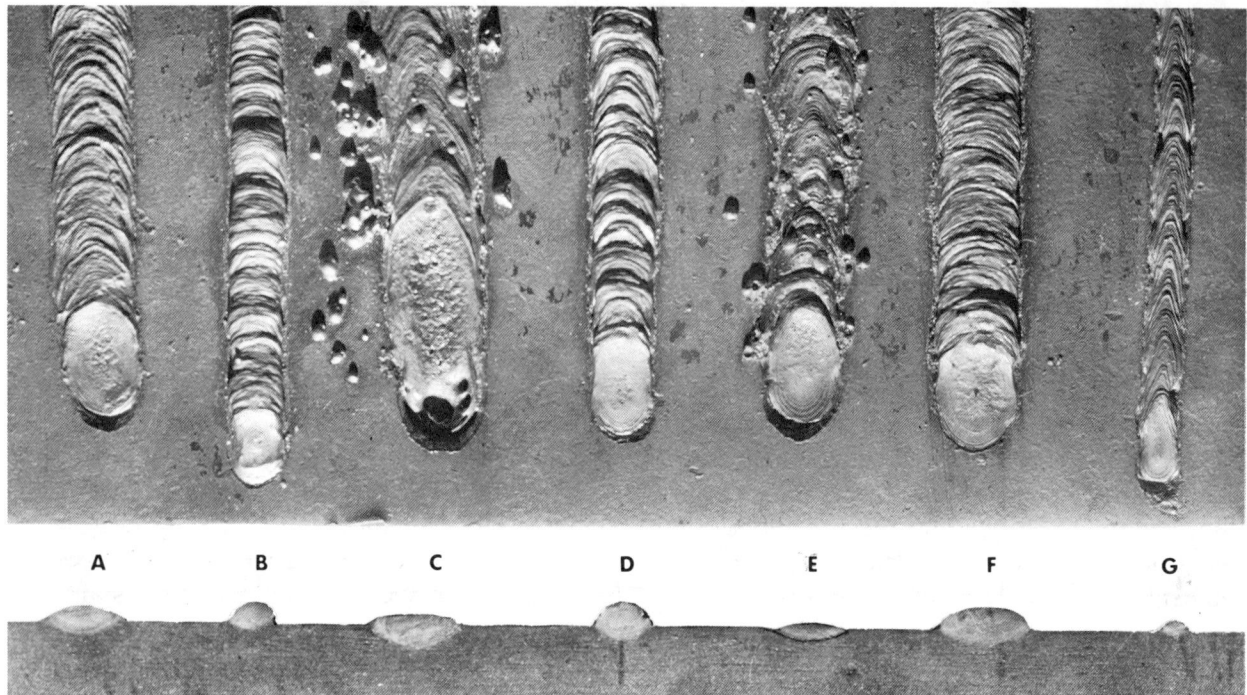

Fig. 6-3. Examples of properly and improperly formed beads. (The Lincoln Electric Co.)
A. Current, voltage, and speed normal
B. Current too high
C. Current too high
D. Voltage too low
E. Voltage too high
F. Speed too slow
G. Speed too fast

large and irregular. When the current is too low, there is not enough heat to melt the base metal and the molten pool will be too small. The result is not only poor fusion, but the beads will pile up and be irregular in shape. See Fig. 6-3.

Correct travel speed. Where the speed is too fast, the molten pool does not last long enough and impurities are locked in the weld. The bead is narrow and the ripples pointed. If the rate of travel is too slow, the metal piles up excessively and the bead is high and wide with straight ripples as illustrated in Fig. 6-3.

Correct electrode angle. The angle at which the electrode is held will greatly affect weld bead shape and is particularly important in fillet and deep groove welding. Electrode angle involves two positions—incline and side angles. *Incline angle* is the angle in the line of welding and may vary from 5° to 30° from the vertical, depending on welder preference and welding conditions. *Side angle* is the angle from horizontal, measured at right angles to the line of welding, which normally splits the angle of the weld joint. See Fig. 6-4 for examples.

Ordinarily, a variation of 15° in either direction from the horizontal will not affect weld appearance or quality. However, whenever undercuts occur in the vertical member of a fillet weld, the angle of the arc should be lowered and the electrode directed more toward the vertical member. Side angle is especially important in multiple pass welding. See Fig. 6-4.

Crater Formation

As the arc comes in contact with the base metal, a pool, or pocket, is formed known as a *crater.* The size and depth of a crater indicate the

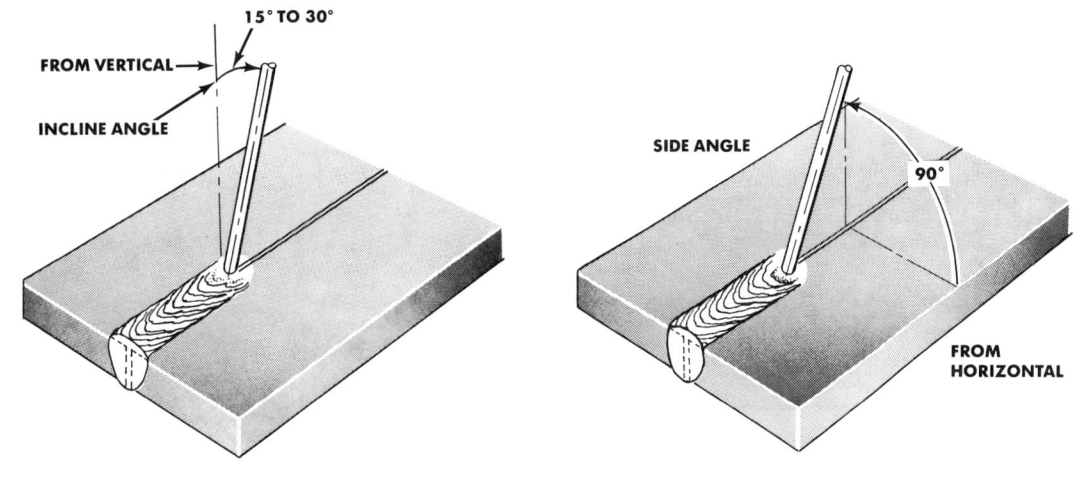

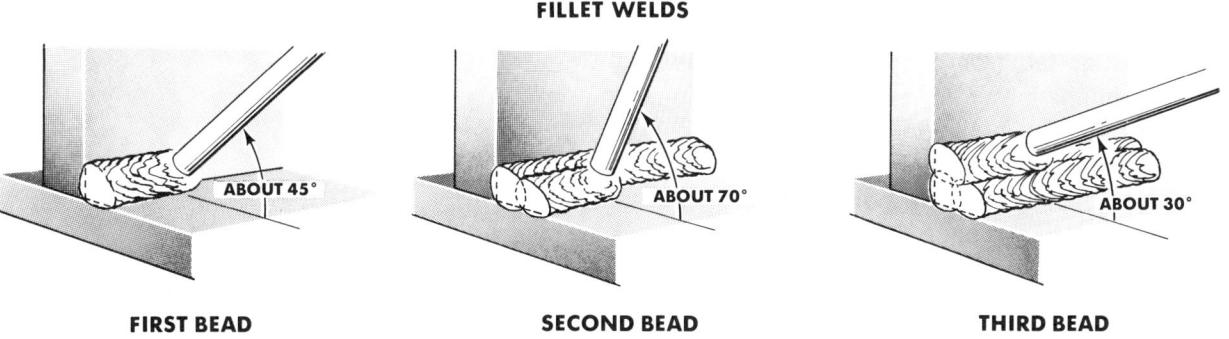

Fig. 6-4. Correct electrode angle is important to make good welds. (Republic Steel Corp.)

degree of penetration. In general, the depth of penetration should be from one-third to one-half the total thickness of the bead, depending upon the size of the electrode, as pictured in Fig. 6-5.

To secure a sound weld, the metal deposited from the electrode must fuse completely with the base metal. Fusion will result only when the base metal has been heated to a liquid state and the molten metal from the electrode readily flows into it. Thus, if the arc is too short, there will be insufficient *spread* of heat to form the correct size crater of molten metal. When the arc is too long, the heat will not be centralized or intense enough to form the desired crater.

Remelting and controlling a crater. An improperly filled crater may cause a weld to fail when a load is applied on a welded structure. Therefore, be sure to fill a crater properly. There is always a tendency when starting an electrode for a large globule of metal to fall on the surface of the plate with little or no penetration. This is especially true when beginning a new electrode

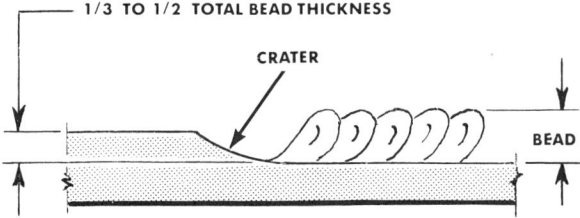

Fig. 6-5. This crater and bead show proper penetration.

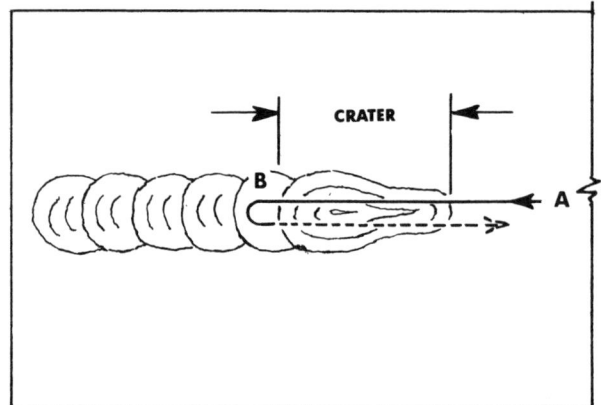

Fig. 6-6. Fill the crater by moving the electrode from A to B, and then back again through the crater to A.

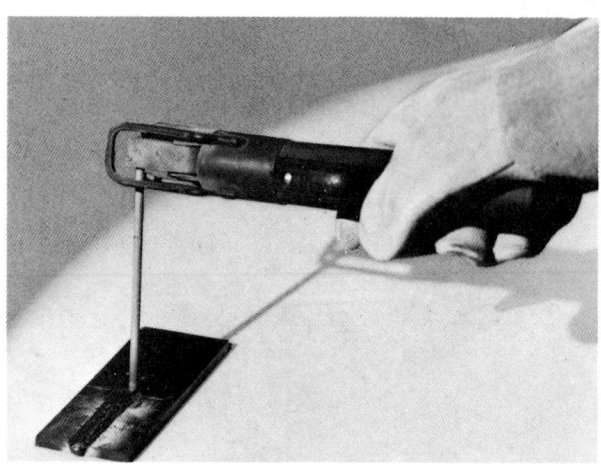

at the crater left from a previously deposited weld. To fill the crater and secure proper fusion, strike the arc approximately 1/2" in front of the crater as shown at *A* in Fig. 6-6. Then bring it back through the crater to point *B* just beyond the crater, and weld back through the crater.

Occasionally, you will find that the crater is getting too hot and the fluid metal has a tendency to run. When this happens, lift the electrode slightly and quickly, and shift it to the side or ahead of the crater. Such a movement reduces the heat, allows the crater to solidify momentarily, and stops the deposit of metal from the electrode. Then return the electrode to the crater and shorten the arc.

Running Continuous Beads

Hold the electrode in a vertical position or slant it slightly away from you as shown in Fig. 6-7.

Move the electrode just rapidly enough so deposited metal has time to penetrate into the base plate. If the current is set properly and the arc is maintained at the correct length, there will be a continuous crackling or frying noise. Learn to recognize this sound. An arc that is too long will have somewhat of a humming sound. Too short of an arc makes a popping sound. Notice the action of the molten puddle and how the

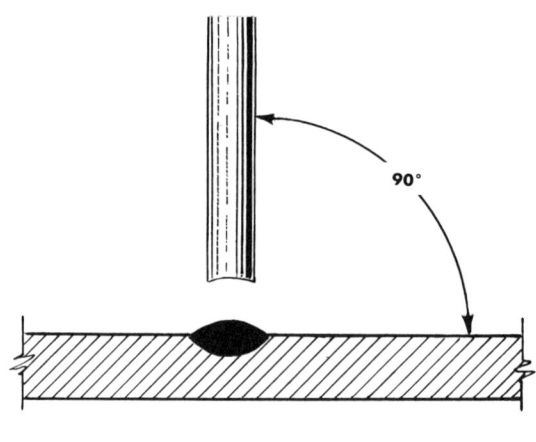

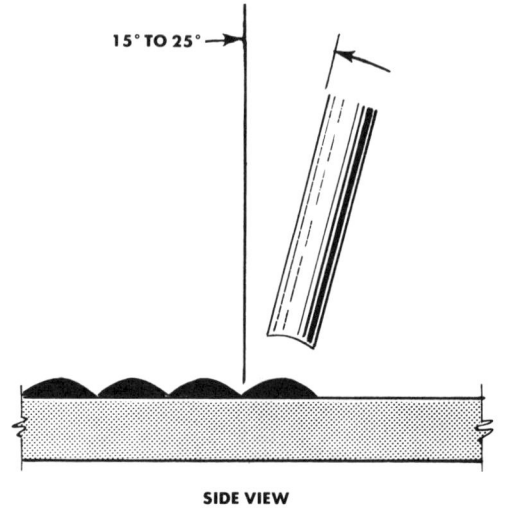

Fig. 6-7. Hold the electrode in this position for running straight beads.

trailing edge of the puddle solidifies as the electrode travels forward.

The appearance of the puddle is often an indication of how good a weld is being made. If the molten metal is clear and bright it means that no molten slag is mixing with the puddle. Slag is brittle and when it flows in the molten metal the weld is weakened. Normally, if the edges of the weld bead have a dull, irregular appearance, it means that slag is being trapped into the puddle.

Checking the Welding Heat

After you have become accustomed to striking an arc and running short weld beads, vary the welding current to see how it affects the welding heat. First turn the machine down about five amperes and check if there is any difference when you run a bead. Then turn it down another five amperes and again try to run a bead. As you reduce the amperage, it soon becomes apparent that there is insufficient heat to melt the base metal. Furthermore, you will find that as the electrode burns off, it does not fuse with the base metal but lies on the surface as spatter which easily scrapes off.

Now reverse the process by gradually raising the amperage. Turn the machine up five amperes in several steps and each time run short beads. It will soon become obvious that as the amperage is increased the arc gets hotter and the electrode melts faster.

From this experiment you can appreciate the importance of having the correct welding heat to make a sound weld. However, as you gain experience in welding, proper adjustment of welding current becomes relatively easy.

Undercutting and Overlapping

Undercutting is a condition that results when the welding current is too high. The excessive current leaves a groove in the base metal along both sides of the bead which greatly reduces the strength of a weld. See Fig. 6-8. Undercutting may also occur when there is insufficient deposition of metal on the vertical plate. This can be corrected by slightly changing the electrode angle.

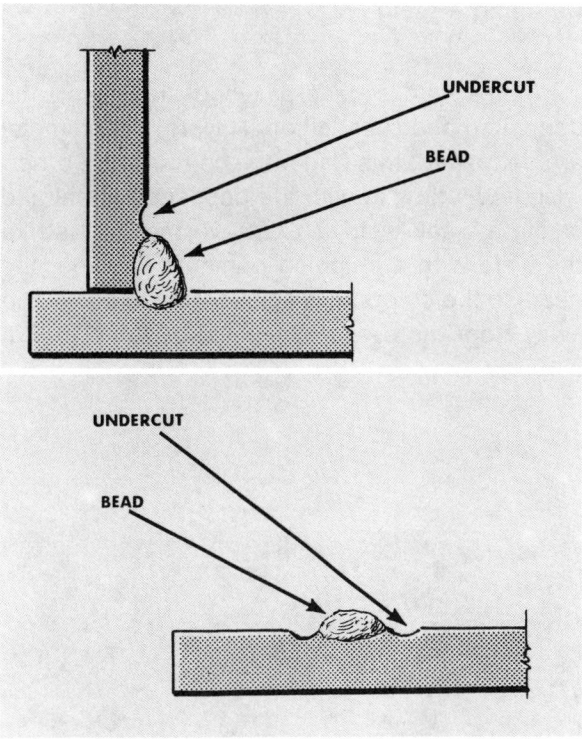

Fig. 6-8. Undercutting greatly weakens a weld.

Overlapping occurs when the current is set too low. In this instance, the molten metal falls from the electrode without actually fusing with the base metal as shown in Fig. 6-9.

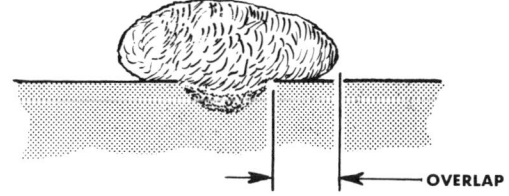

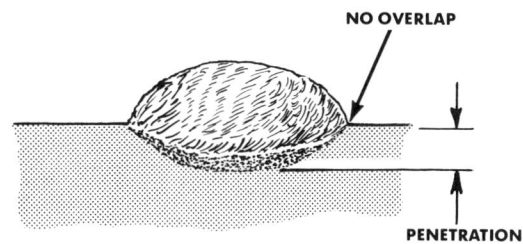

Fig. 6-9. When not enough heat is used, overlapping occurs as shown in top illustration. A satisfactory weld is shown at the bottom.

Cleaning a Weld

When a weld is made, a layer of slag covers the deposited bead. If additional layers of weld metal are deposited, this slag must be removed; otherwise it will flow in with the deposited metal and cause a weak weld. To remove the slag, strike the weld with a chipping hammer. Hammer the bead so the chipping is away from the body, and away from the eyes and face as pictured in Fig. 6-10.

Fig. 6-10. Slag is removed from the weld by chipping.

Fig. 6-11. After chipping, brush the weld with a wire brush. (Hobart Brothers Co.)

CAUTION: Always wear eye protectors when chipping. Do not pound the bead too hard; otherwise the structure of the weld may be damaged. After the slag is loosened, drag the pointed end of the hammer along the weld where it joins the plate. This will remove the remaining particles of slag. Follow the chipping with a good, hard brushing, using a stiff wire brush as illustrated in Fig. 6-11.

Weaving the Electrode

Weaving is a technique used to increase the width and volume of the bead. Enlarging the size of the bead is often necessary on deep groove or fillet welds where a number of passes must be made. Fig. 6-12 illustrates several weaving patterns. The pattern used depends to some extent on the position of the weld. In subsequent units additional instructions will be given as to the most suitable weaving pattern for specific weld joints.

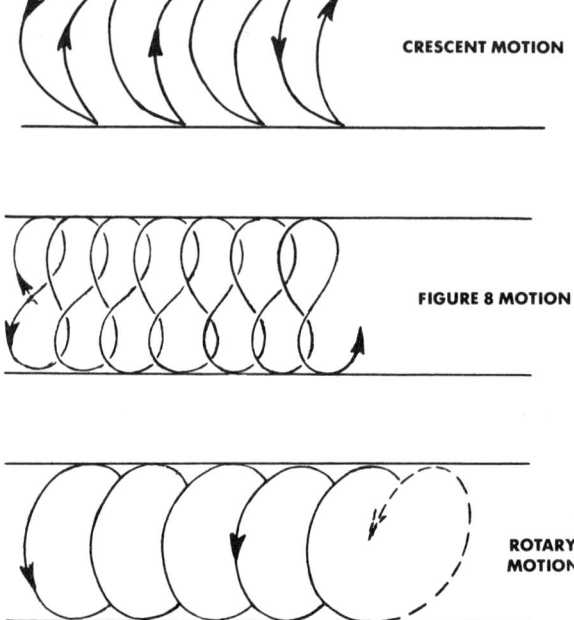

Fig. 6-12. The crescent, figure 8, and rotary motions are three typical weaving patterns.

SELF QUIZ

Correct answers are listed in the back of the book.

Fill-In

Supply the missing word or words where blanks appear.

1. A good arc will make a _____ kind of sound.
2. When an arc is too long, large _____ of metal melt off the electrode and drop on the plate surface.
3. Too long of an arc will produce an irregular bead without sufficient _____.
4. If an arc is too short, the electrode _____ frequently and the beads are _____ and irregular.
5. The length of the arc should be approximately equal to the _____ of the electrode.
6. A too long arc allows the _____ to contaminate the weld.
7. If the current is too high, the electrode _____ too fast.
8. When the current is too low, there is not enough _____ to produce good penetration.
9. High travel speed produces _____, pointed ripples.
10. If the travel speed is too slow, the beads will usually be too _____ and _____.
11. The size and depth of a crater will indicate the amount of _____.
12. The size and depth of a crater are governed by the _____ length.
13. When a crater gets too hot and the metal has a tendency to run, _____ the electrode slightly and quickly, and move it to the _____ or _____ of the crater.
14. When reaching the end of a weld the crater must be completely _____.
15. Undercutting is a condition that results from a too _____ current.
16. Overlapping occurs when the _____ is too _____.

WELDING ASSIGNMENT

Plate no. 1. After you have mastered the operation of striking an arc, secure a steel plate 1/4" × 4" × 6". With a soapstone, draw a number of lines approximately 3/4" apart, as shown in Fig. 6-13. Use a 1/8" diameter E-6010 or E-6011 electrode and run continuous beads on these lines, starting from the left edge and working to the right. After the plate is filled, remove the slag and examine the beads.

Plate no. 2. This plate will give you an opportunity to develop skill in re-striking an arc while making a continuous bead. Draw a series of straight lines on the plate and divide these lines

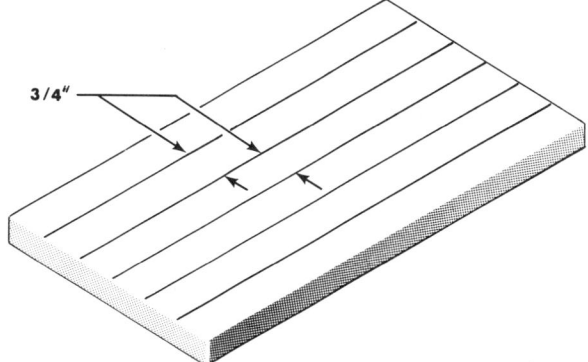

Fig. 6-13. Using a steel plate with lines 3/4" apart, deposit continuous beads from left to right.

58 Arc Welding

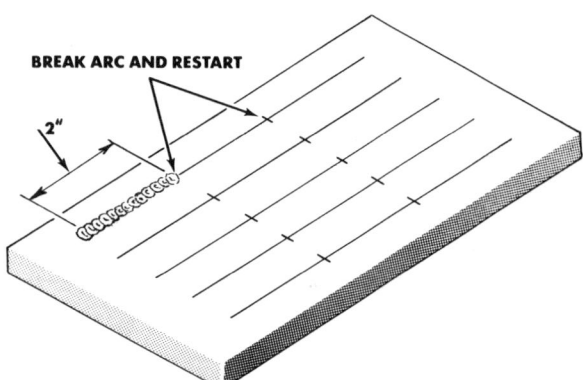

Fig. 6-14. To develop skill in re-starting an arc, break the arc and re-start every two inches.

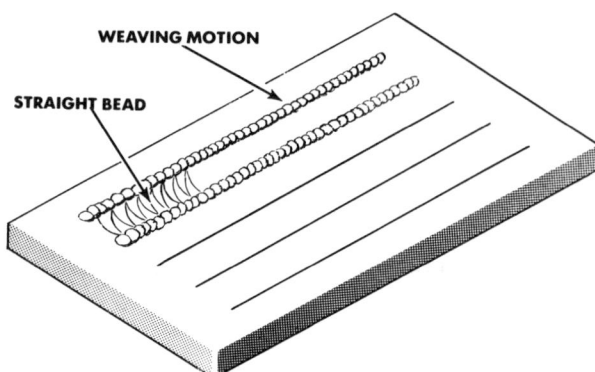

Fig. 6-15. Practice weaving by depositing a weld from left to right between the straight beads.

into 2-inch sections as shown in Fig. 6-14. Run a bead over the first line but break the arc when reaching the end of the 2-inch mark. Re-start the arc and continue the deposit for another two inches; then repeat the practice of breaking the arc and refilling the crater. Follow this procedure until skill is mastered in depositing uniform and continuous beads with properly filled craters.

Plate no. 3. Lay out a series of straight lines on a 1/4" × 4" × 6" plate. Run continuous beads over these lines and then clean each bead. Now proceed to practice weaving by depositing a weld back and forth between the first pair of continuous beads as illustrated in Fig. 6-15. Use one type of weaving motion to fill the first space and then try several other weaves on the remaining sections. Make certain that the short beads are fused into the long, straight beads. Continue the weaving practice on several plates until a workman-like job is accomplished.

Plate no. 4, depositing a weld metal pad. *Padding* or surfacing is a process for building worn surfaces of shafts, wheels, and other machine parts. The operation consists of depositing several layers of beads, one on top of the other. See Fig. 6-16. Fill the section completely with

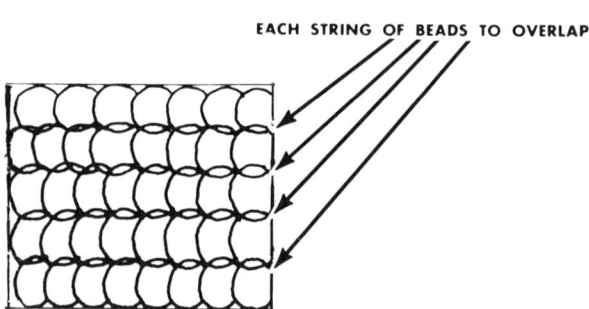

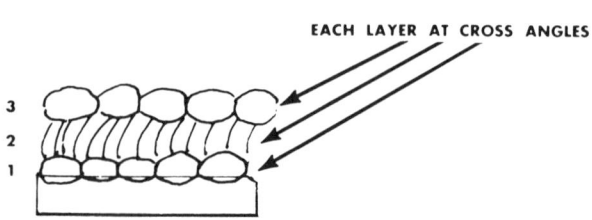

Fig. 6-16. Practice plate for building a pad.

beads, clean the weld thoroughly, and then deposit a second layer of weaving beads about 1/2" wide at right angles to the first layer. After the weld is cleaned, deposit a third layer at right angles to the second layer.

UNIT 7
Shielded Metal Arc—Flat Position Welding

Although welding can be done in any position, the operation is simplified if the joint is flat. When placed in this way, the welding speed is increased, the molten metal has less tendency to run, better penetration can be secured, and the work is less fatiguing.

JOINT DESIGN

Just what type of joint is best suited for a particular job depends on many factors. Although the designer or engineer is primarily responsible for determining the kind of joint to be used, if a welder knows something about joint design he usually produces welds that will better meet the established specifications for the job. In general the following factors should be considered in preparing a joint for welding:[1]

1. *Fit-up must be consistent for the entire joint.* Sheet metal and most fillet and lap joints should be clamped tight for their entire length. Gaps or bevels must be accurately controlled over the entire joint. Any variation in a given joint will force the operator to slow the welding speed to avoid burn through and to handle the different electrode manipulation required by the fit-up variation.

2. *Sufficient bevel is required for good bead shape and penetration.* See Fig. 7-1. Insufficient bevel prevents getting the electrode into the joint. A deep, narrow bead may lack penetration and may have a strong tendency to crack.

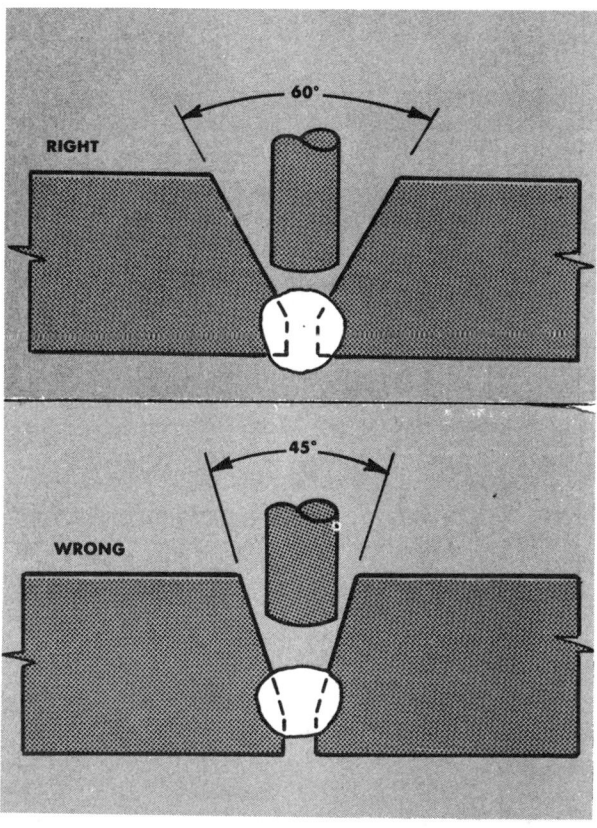

Fig. 7-1. Correct groove bevel is essential for a good weld.

---
1. The Lincoln Electric Co.

60 Arc Welding

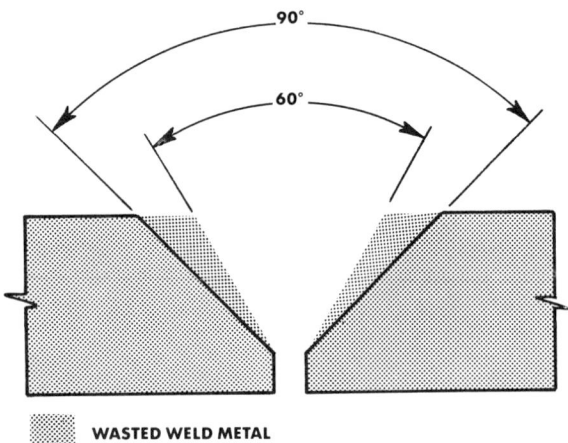

Fig. 7-2. Grooves that are too wide usually result in greater welding costs.

3. *Excess bevel wastes weld metal.* See Fig. 7-2. Since filler metal in the form of electrodes and wire is expensive, any variation from the recommended groove angle size simply contributes to the cost of a weld both in terms of material and time.

4. *Sufficient gap is needed for full penetration.* See Fig. 7-3. Unless there is adequate penetration, a welded joint will not withstand the loads imposed on it. Although proper penetration depends to some extent on electrode manipulation, the first essential is providing a correct root opening.

5. *Either a ⅛" [0.125"] land (root face) or a back-up strip is required for fast welding and good quality.* See Fig. 7-4. Feather-edge preparations require a slow costly seal (root) bead.

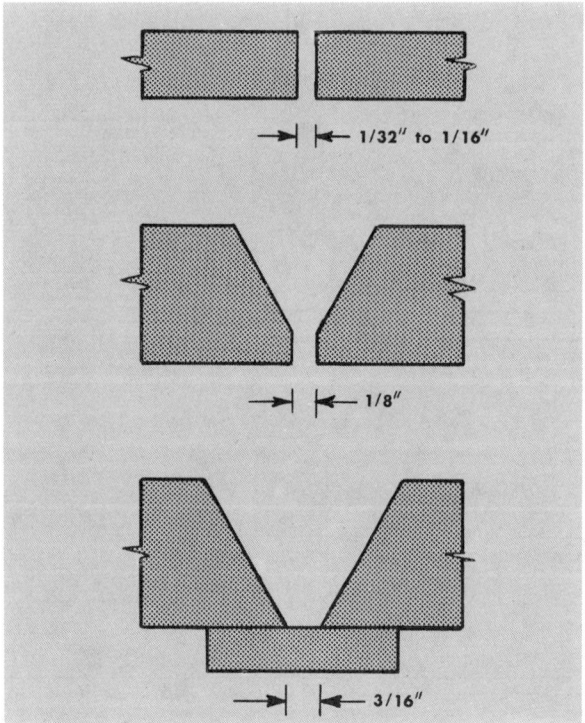

Fig. 7-3. To make a sound weld a correct root opening is important.

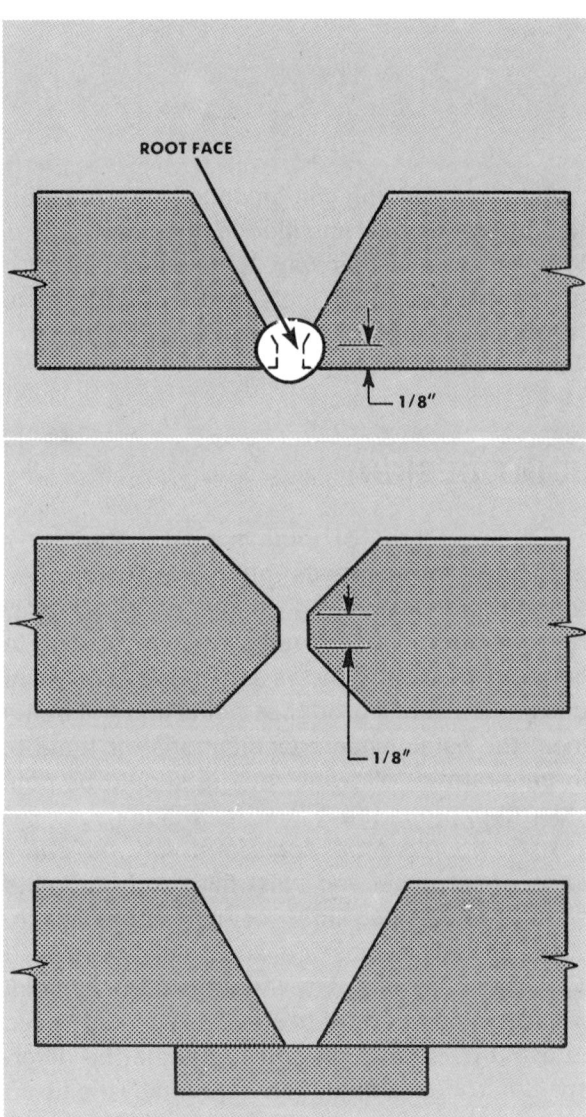

Fig. 7-4. A proper root face should be provided for quality welds.

However, double-V butt joints without a land are practical when the root bead is offset by easier edge preparations and the gap can be limited to about 3/32".

WELD PASSES

In carrying out some welding operations, very often the pieces have to be tack-welded. Tack welds are simply short sections of weld beads 1/4" to 1/2" long used to maintain the proper root opening between the two sections of metal being welded. See Unit 3. These tack welds are spaced along the seam and must be firmly fused into the joint.

Once the joint is tacked, the remaining weld passes are deposited. The first pass, known as the *root pass* or *stringer bead,* is a narrow bead laid in the bottom of the V. It is made with a small diameter electrode by moving it straight down into the groove without any weaving motion. See Fig. 7-5. Its principal function is to fill the root opening and join the two metal sections. Since it serves as the base for the other passes, it is very important that it produces complete penetration. Complete penetration is more assured if the root pass penetrates the bottom surface of the groove, but not more than 1/16", and fuses all tack welds previously made.

The next layer is called the *fill* or *filler pass.* One or more filler passes may be needed to fill the groove, depending on the thickness of the metal. In depositing the filler passes a slight weaving motion is generally advisable to insure proper fusion to the previously laid beads and sides of the groove joint.

The final pass, called the *cap* or *cover pass* (sometimes referred to as the wash bead), is intended to provide additional reinforcement to the weld and give it a nice appearance. The cover pass should not extend more than 1/16" above the plate surface. Inasmuch as it has to extend completely over and beyond the filler passes, a weaving motion is necessary to cover the wider area.

How to Weld a Butt Joint

The butt joint is often used when structural pieces have flat surfaces, as in tanks, boilers, and a variety of machine parts. The joint may be opened, closed, or the edges beveled, as in Fig. 7-6. A *closed butt joint* has the edges of the two plates in direct contact with each other. This

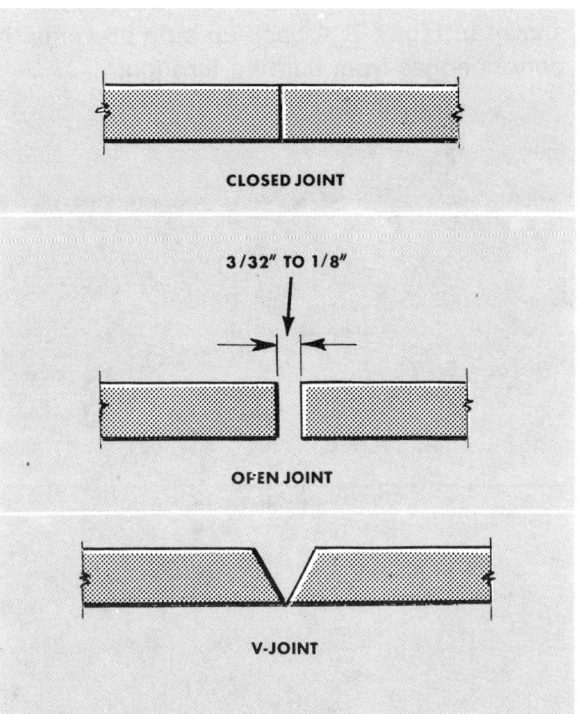

Fig. 7-6. There are three types of butt joints.

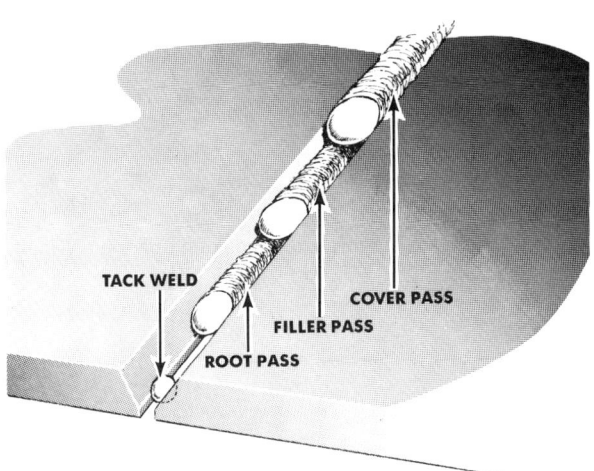

Fig. 7-5. Types of weld passes.

62 Arc Welding

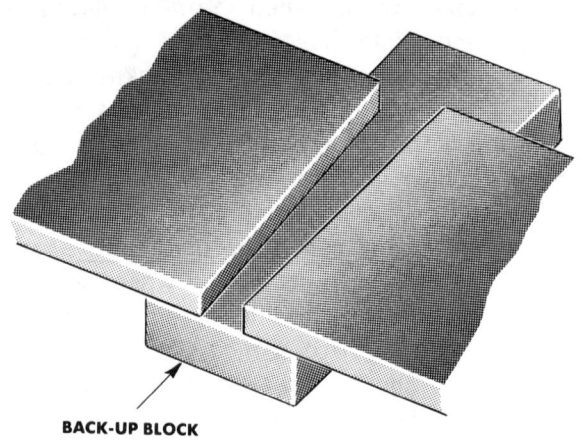

Fig. 7-7. A back-up strip or block should be used on an open-butt joint.

joint is suitable for welding steel plates that *do not exceed 1/8″ to 3/16″ in thickness.* Heavier metal can be welded but only if the machine has sufficient amperage capacity and if heavier electrodes are used. *Remember that on thick metal it is difficult to secure ample penetration to produce a strong weld by a single pass bead.*

In the *open butt joint* the edges are placed slightly apart, usually 3/32″ to 1/8″, to allow for expansion. As a rule, a back-up strip or block of scrap steel or copper is placed under this joint as shown in Fig. 7-7. A back-up strip prevents the bottom edges from burning through.

When the thickness of the metal exceeds 1/8″ the edges of a butt joint should be beveled. The beveling can be done by cutting the edges with a flame torch or by grinding them on an emery wheel. The included angle of the V should not exceed 60°, to limit the amount of contraction that usually results when the metal cools. The edges may be shaped in several ways as shown in Fig. 7-8. Notice that on heavy metal 3/8″ or more in thickness the edges are beveled on both sides. Beveling in this manner insures better penetration, requires less weld metal, and contraction forces are better equalized.

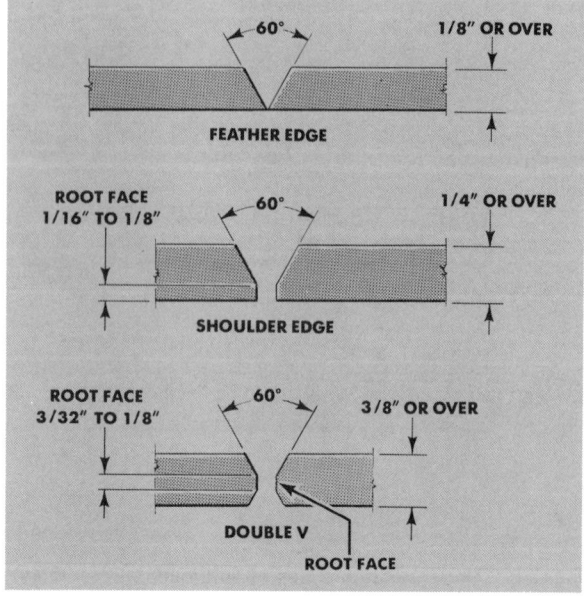

Fig. 7-8. Here are the three methods of preparing a V-joint.

Shielded Metal Arc—Flat Position Welding

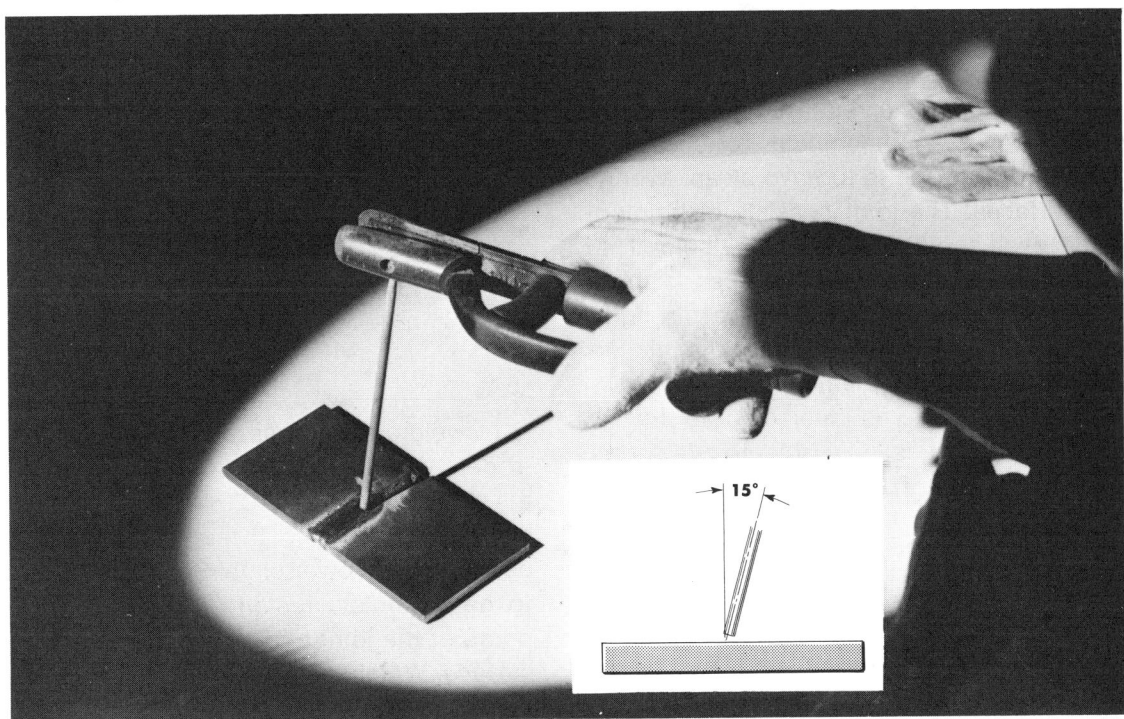

Fig. 7-9. Position of the electrode for welding a butt joint. (The Lincoln Electric Co.)

Welding procedure. Fig. 7-9 illustrates the position of the electrode for welding a butt joint. However, when the joint consists of two pieces of different thicknesses, adjust the position of the electrode so the greatest portion of the heat is concentrated on the thickest plate.

In welding thin stock with a single pass, as in a closed or open butt joint, simply allow the electrode to travel along the seam without any weaving motion. Move the electrode slow enough to give the arc sufficient time to melt the metal. Take care that the travel is not too slow, because the arc will burn through the metal.

When a multiple pass is to be made in a grooved joint, be sure to hold the electrode down in the groove so it almost touches both sides of the joint while depositing the root pass. Move the electrode fast enough to keep the slag flowing back on the finished weld. If the electrode is not moved rapidly enough, the slag may become trapped in the bottom of the weld, thereby preventing proper fusion.

After completing the root pass, proceed with whatever filler passes are required and then complete the weld with a cover pass. See Fig. 7-10. Remember, always remove the slag completely after each pass. If any slag particles are allowed to remain, they will weaken the weld.

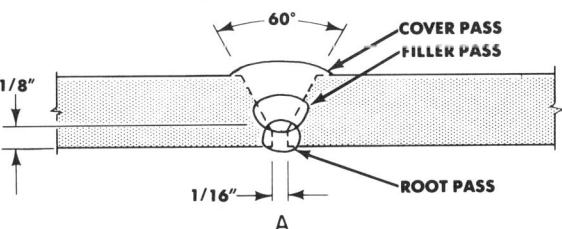

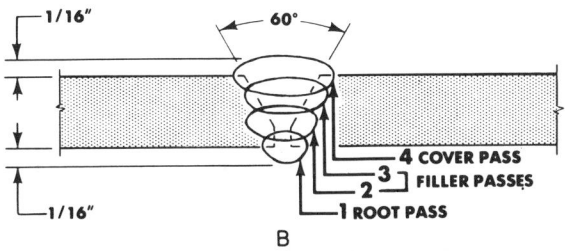

Fig. 7-10. Sequence of passes in a multiple-pass butt joint.

Making a Single-Pass Fillet Lap Weld

The lap joint is one of the most frequently used joints in welding. It is a relatively simple joint, since no beveling or machining is necessary. One standard requisite is to have clean, evenly aligned surfaces. The joint consists of lapping one edge over another and joining them. The amount the edges should overlap depends upon the thickness of the plates and the strength required of the welded piece. Usually the thicker the plates, the greater the amount of overlap.

When the structure is subjected to heavy bending stresses, it is advisable to weld the edges of both sides of the joint as illustrated in Fig. 7-11.

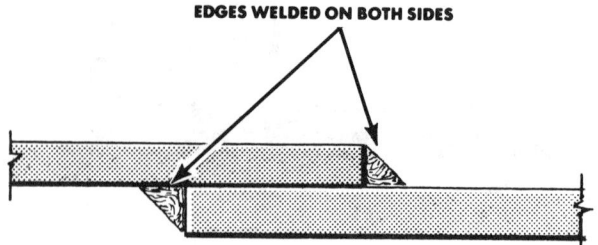

Fig. 7-11. For a strong joint weld both edges.

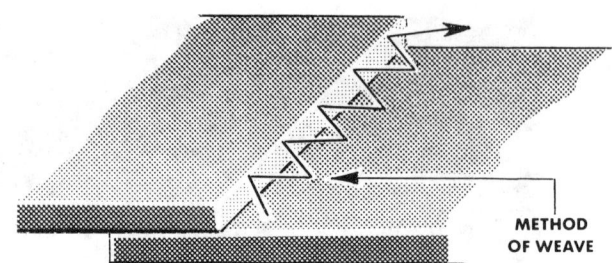

Fig. 7-13. Use this weave for a flat lap weld.

To weld a single-pass lap joint, hold the electrode at a 45° angle and point it toward the weld as shown in Fig. 7-12. Weave the electrode slightly, maintaining the arc for a little longer period on the lower plate. See Fig. 7-13. Be sure to get complete fusion at the *root*, or joining point, of the joint and avoid overlaps on the top surface. Watch the crater carefully to prevent an undercut on the bottom plate. Fig. 7-14, *top,* illustrates a *properly made fillet weld* on a lap joint. A weld made as in Fig. 7-14, *center,* usually

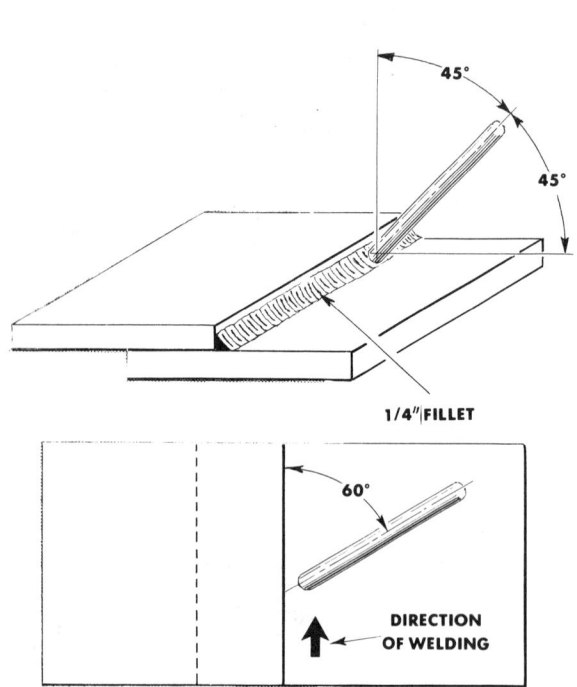

Fig. 7-12. For a sound lap weld, hold the electrode as shown in the above two views.

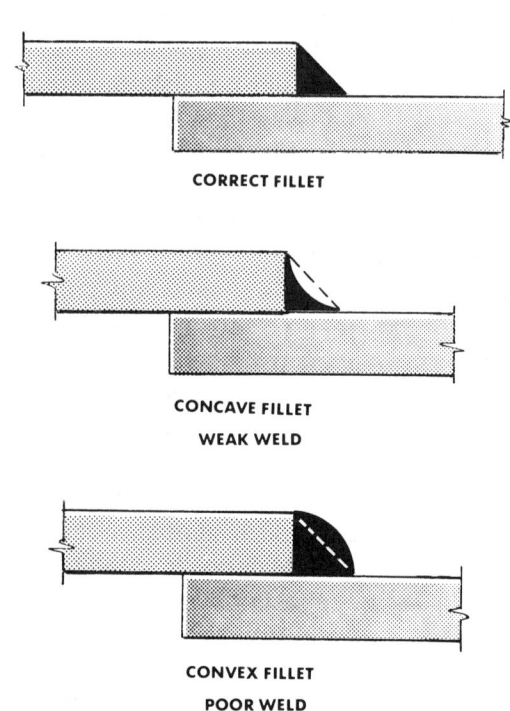

Fig. 7-14. How a good lap weld and two poor lap welds appear from a side view.

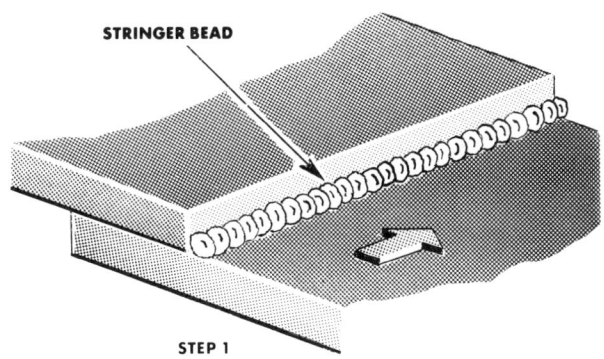
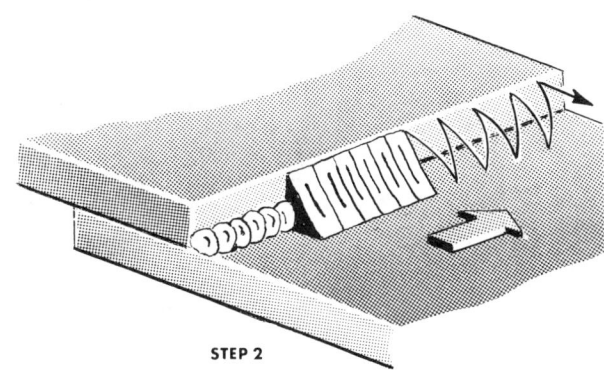

Fig. 7-15. Making a multiple-pass fillet weld on a lap joint.

is too weak because it lacks sufficient reinforcing material, while the weld shown in Fig. 7-14, *bottom,* has too much waste metal, which is of no value to the joint.

Making a Multiple-Pass Fillet Lap Weld

When an exceptionally strong lap joint is required, especially on heavy plates 3/8" and over in thickness, a multiple-pass fillet weld is recommended. This joint has two or more layers of beads along the seam, with each bead lapping over the other.

To make such a weld, deposit the first bead as shown in Fig. 7-15 by moving the electrode straight down the seam without any weaving motion. Clean the weld carefully and lay the second pass over this *stringer bead*. During the second pass, weave the electrode, pausing for an instant at the top of the weave to favor or deposit extra metal on the vertical edge of the upper plate.

Making a Single-Pass T-Fillet Joint

The T-fillet joint is frequently used in fabricating straight and rolled shapes. See Fig. 7-16. The strength of this joint depends considerably on having the edges of the joint fit close together. The T-joint should not be used if it is subjected to heavy stresses from the opposite direction of the welded seam. This weakness can be partially overcome by using a double fillet—that is, welding both sides of the joint. See Fig. 7-17.

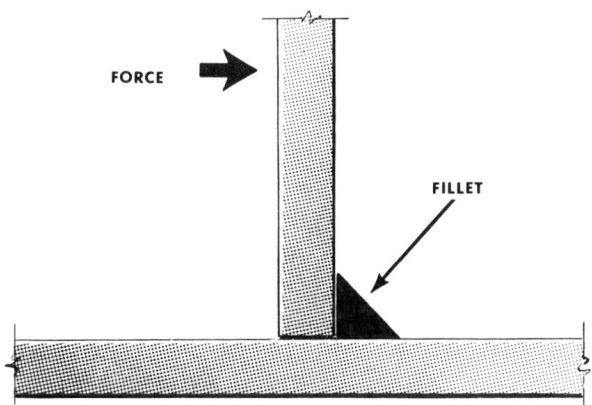

Fig. 7-16. This joint would be weak if subjected to stresses from the direction indicated by the arrow.

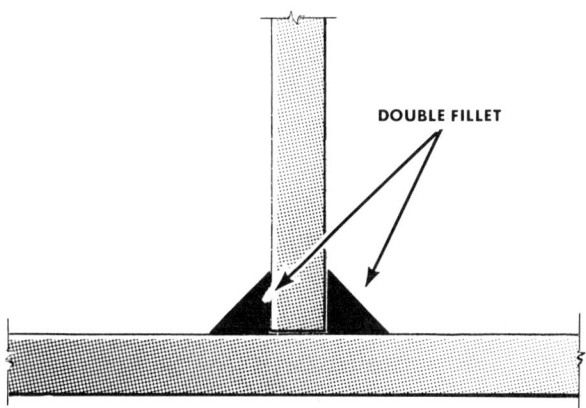

Fig. 7-17. A double fillet T-joint is stronger.

66 Arc Welding

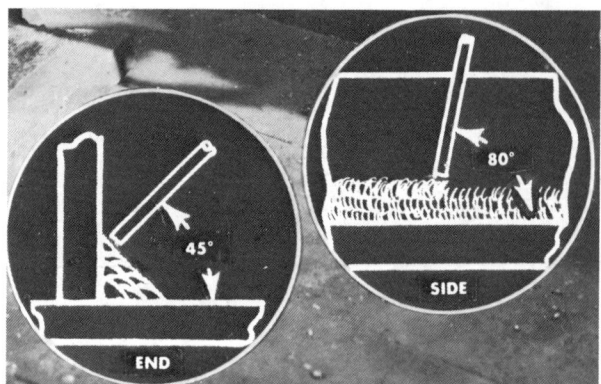

Fig. 7-18. Position of the electrode for welding a T-fillet joint.

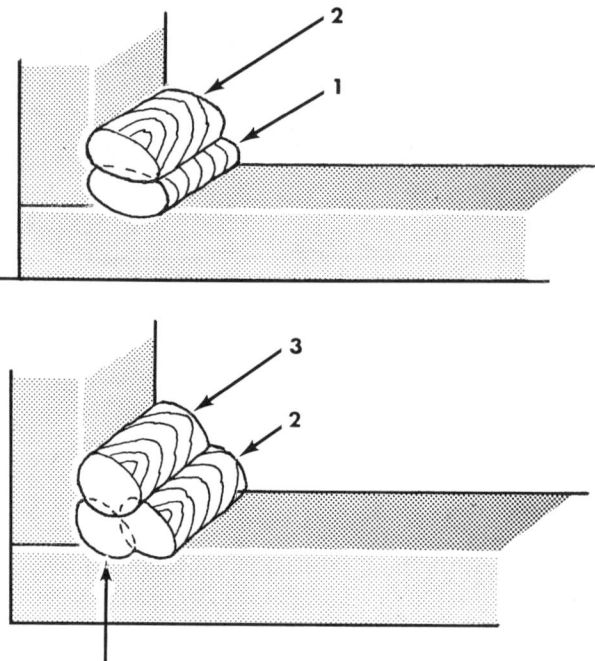

Fig. 7-19. Sequences of passes for a multiple T-fillet weld.

To weld a T-fillet joint, hold the electrode as shown in Fig. 7-18 and advance it in a straight line without any weaving motion. Point the tip of the electrode toward the completed portion of the weld and travel rapidly enough to stay ahead of the molten pool. Concentrate the arc more on the lower plate to prevent undercutting the upper plate. Watch the crater closely so it will form a bead with the correct contour.

Making a Multiple-Pass T-Fillet Joint

When a very strong T-joint is required, make a wider fillet along the seam. You can get a wider fillet by running several layers of beads as illustrated in Fig. 7-19. Deposit the first bead as described in making a single pass T-fillet joint. Remove the slag and lay the second bead over the first, weaving the electrode sufficiently to secure the desired width fillet. Deposit additional layers if necessary to get the right size fillet, but *be sure to clean off the slag after each pass.*

How to Make an Outside Corner Weld

The outside corner weld, as shown in Fig. 7-20, is often used in constructing rectangular shaped objects such as tanks, metal furniture, and other machine sections where the outside corner must have a smooth radius.

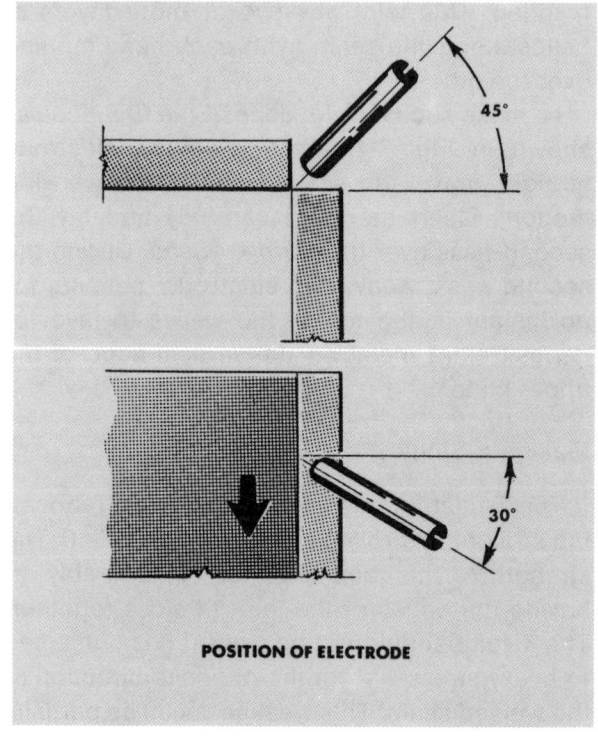

Fig. 7-20. An outside corner weld. The angle of the electrode is viewed from the side (top), and overhead (bottom).

To make an outside corner weld, tack the two plates and run a bead along the edge with the electrode held as indicated in Fig. 7-20. On light stock, one bead is usually enough. Heavy stock will probably require a series of passes to fill in the corner.

SELF QUIZ

Correct answers are listed in the back of the book.

Identification

Place the answers on the blanks.

1. Identify the joints shown below.

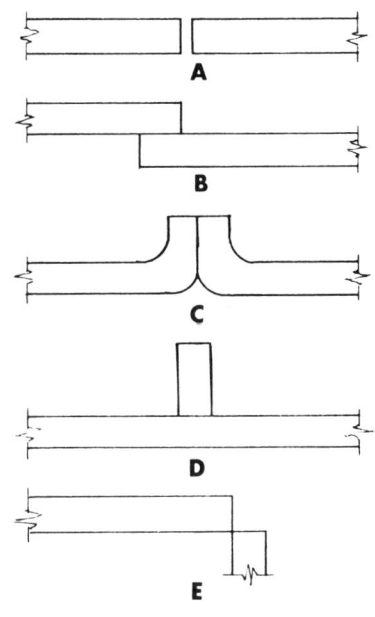

A _____
B _____
C _____
D _____
E _____

2. Identify the following butt joints.

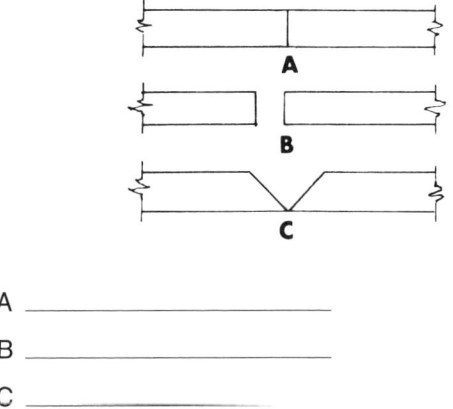

A _____
B _____
C _____

Multiple Choice

Circle the letter which represents the correct answer.

3. A joint made with a single pass is
   a. A poor weld
   b. an exceptionally strong weld
   c. often satisfactory for some jobs
   d. not used to get adequate penetration
4. A tack weld is generally used to
   a. secure better fusion
   b. hold sections in position for welding
   c. increase the deposition rate
   d. prevent excessive distortion

5. A lap joint is frequently welded on both sides to
   a. prevent distortion
   b. minimize distortion
   c. provide greater fusion area
   d. offset heavy bending stresses
6. In welding a lap joint, the electrode should be
   a. held so it bisects the edges at a 45° angle
   b. slanted slightly away from the weld
   c. kept buried deep in the groove
   d. moved rapidly along the edges
7. Which of the following lap welds is considered to be the best?

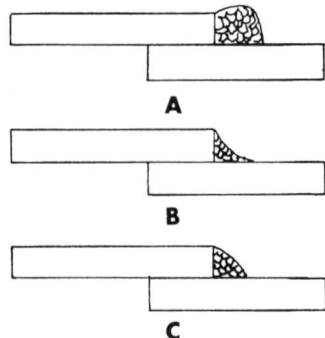

   a.
   b.
   c.
8. A T-joint welded on one side only
   a. should not be used
   b. is acceptable if not subjected to heavy stresses from the opposite side
   c. is acceptable if made with a weaving motion
   d. has expansion and contraction forces difficult to control
9. A closed butt joint is satisfactory provided it is made
   a. with large diameter electrodes
   b. with the edges beveled
   c. without any requirements for deep penetration
   d. on plates that do not exceed 1/8" in thickness
10. A back-up strip is sometimes used in welding an open butt joint
    a. to secure better penetration
    b. if the metal exceeds 1/2" in thickness
    c. to prevent the bottom edge from burning through
    d. to prevent excessive expansion forces
11. Which of these factors does *not* apply to multiple-pass welds?
    a. Fit-up must be the same throughout the entire joint.
    b. Sufficient bevel is necessary for good bead shape and penetration.
    c. Sufficient root opening is important for adequate penetration in thick plates.
    d. Root face is necessary on all lap joints.
12. Groove openings for multiple-pass welds should have an included angle of about
    a. 90°
    b. 45°
    c. 60°
    d. 30°
13. The number of passes to be made on a joint depends on
    a. joint design
    b. thickness of the metal
    c. included angle of the groove
    d. amperage setting

True and False

Circle the letter T if the statement is true or the letter F if it is false.

14. T F In a multiple-pass weld, the first pass is usually known as a stringer bead.
15. T F A double fillet weld on a T-joint is required when it is subjected to heavy stresses.
16. T F The sequence or placement of each pass on a joint does not affect the strength of the joint itself.
17. T F Multiple-pass welds are not recommended for overhead welding.
18. T F A wash pass is always required in a multiple-pass weld.
19. T F In making a multiple-pass weld on a lap joint, the position of the electrode always remains the same.
20. T F A root pass should be made with a weaving motion.

WELDING ASSIGNMENT

I. Welding a Corner Joint

1. Obtain several pieces of 1/8" or 3/16" low carbon steel plates and select appropriate electrodes.

2. Place the edges of two pieces to form a corner joint and tack weld them. A couple of firebricks will hold the joint in position. See Fig. 7-24.

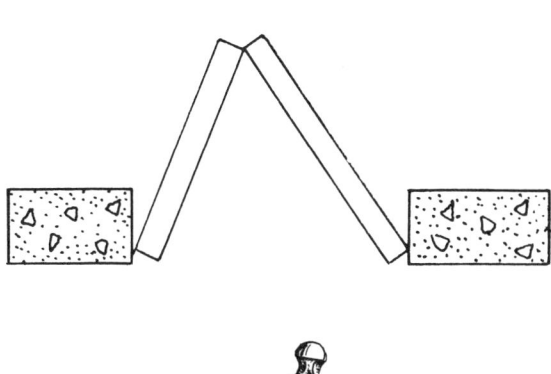

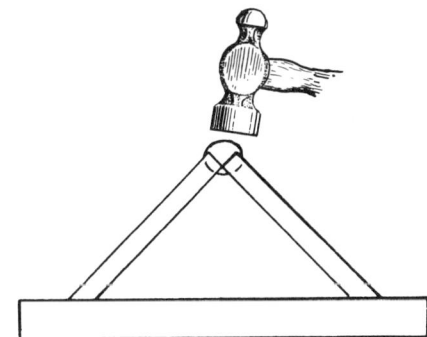

Fig. 7-24. Corner plate practice weld.

3. Run a single pass along the edges without any weaving motion.

4. Test the weld by flattening it as shown in Fig. 7-24. Check your weld results by noting the items in the Weld Analysis Check List shown at the end of this unit.

5. Practice welding this joint until your instructor thinks you can make a sound corner weld.

II. Welding a Butt Joint—Single Pass

1. Use the same kind of metal plates and electrodes you used in the first job.

2. Practice butt welds on both closed and open joints. Tack weld the pieces. Be sure to allow the required spacing on the open-butt joint. You may want to use a back-up strip when welding an open-butt joint.

3. Run a single bead over the joint without any weaving motion. Make certain the crater at the end of the weld is filled completely.

4. Test the weld by clamping it in a vise with the weld parallel to the top of the jaws. Strike the metal with a hammer, bending the weld to 90° or more. See Fig. 7-25. Make one bend (A) with the

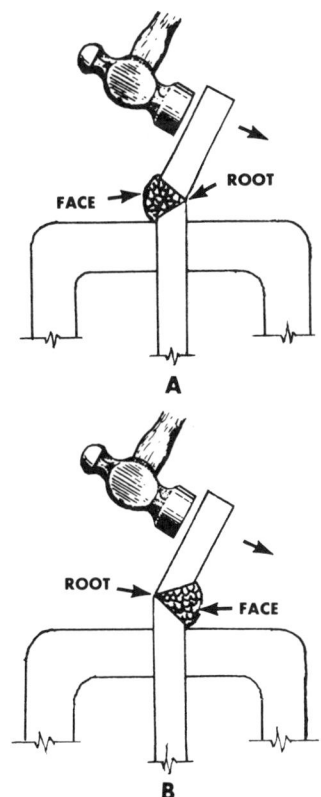

Fig. 7-25. Bend tests for butt joint.

face of the weld in tension and the root in compression. On another test (B), bend the weld so its face is in compression and the root in tension. This latter bend will open up the weld, and you will be able to check the penetration better.

Evaluate your results by using the Weld Analysis Check List shown at the end of this unit. Also compare the welds made on an open and closed joint.

5. Continue to practice making butt welds until your instructor certifies that you have mastered this skill.

III. Welding a Lap Joint—Single Pass

1. Use the same kind of metal plates and electrodes you used in the first two jobs.
2. Arrange the pieces to form a lap joint and tack weld them.
3. Run a single pass without any weaving motion along the edges. Hold the electrode as directed for this joint. Weld on one side only.
4. Test the weld by bending it in the vise as shown in Fig. 7-26. Use the Weld Analysis Check List, shown at the end of this unit, to evaluate the results.
5. Continue to practice making lap welds until your instructor tells you that you have mastered this skill.

IV. Welding a T-Joint—Single Pass

1. Use the same kind of metal plates and electrodes you used in the previous jobs.
2. Tack weld the pieces to form a T-joint.
3. Run a single fillet bead along the edge without any weaving motion. Weld on one side only.
4. Test the weld as shown in Fig. 7-27 and check the results by using the Weld Analysis Check List shown at the end of this unit.
5. Continue to practice welding T-joints until your instructor says that you have mastered this skill.

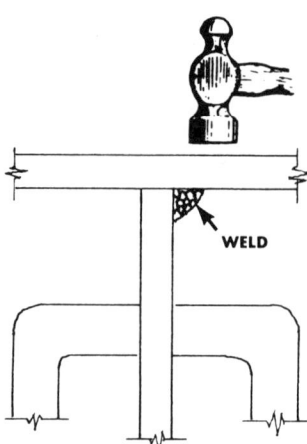

Fig. 7-27. Bend test for T-Joint.

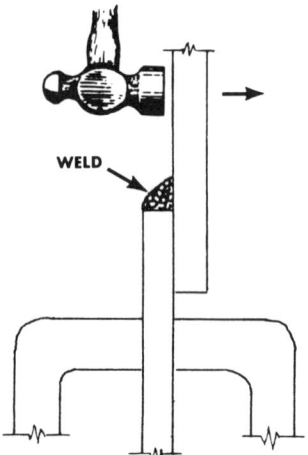

Fig. 7-26. Bend test for lap joint.

V. Making Multiple Pass Welds

1. Obtain several metal plates of a thickness specified by your instructor.
2. Practice multiple-pass welds on lap, T, and butt joints. Be sure to clean the slag off after each pass. Follow the instructions in this book for depositing the various layers of beads.
3. For practice purposes, deposit two passes on lap and butt joints and three passes on T-joints. Weld both sides of the joints. See Fig. 7-28. If stock is available, try welding a double V butt joint.

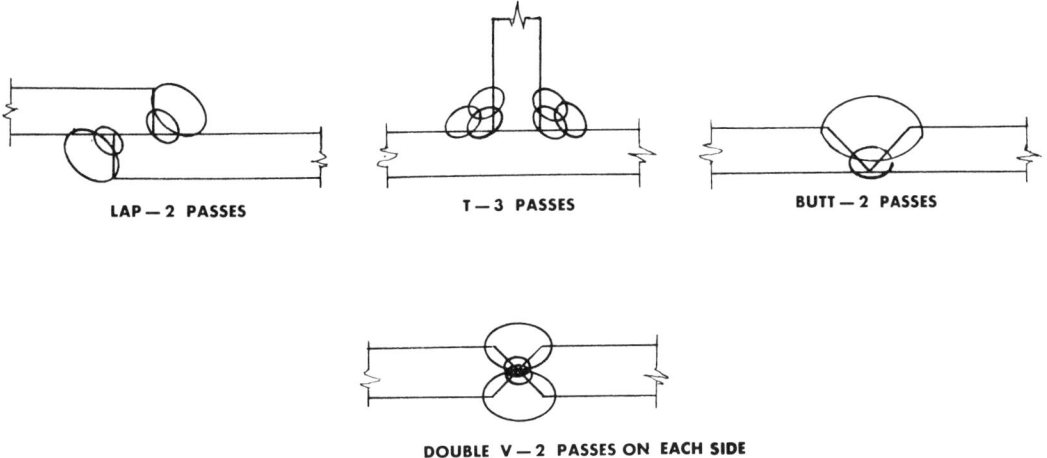

Fig. 7-28. Multiple-pass practice welds in flat position.

Weld Analysis Check List

	Yes	No
1. Bead widths are right size	___	___
2. Beads have uniform ripples	___	___
3. Weld beads are too flat	___	___
4. Weld beads are too high	___	___
5. Weld penetration is insufficient	___	___
6. Weld penetration is excessive	___	___
7. Cold laps on surface	___	___
8. Weld has surface porosity	___	___
9. Weld has subsurface porosity	___	___
10. Weld has crater cracks	___	___
11. Weld has burn thru	___	___
12. End crater is filled	___	___
13. Weld passed bend test without cracking	___	___

UNIT 8
Shielded Metal Arc—Horizontal Welding

On many jobs it is practically impossible to weld pieces in the flat position. Occasionally the welding operation must be done while the work is in a horizontal position. A weld is in a horizontal position when the joint is on a vertical plate and the line of weld runs on a line with the horizon as in Fig. 8-1.

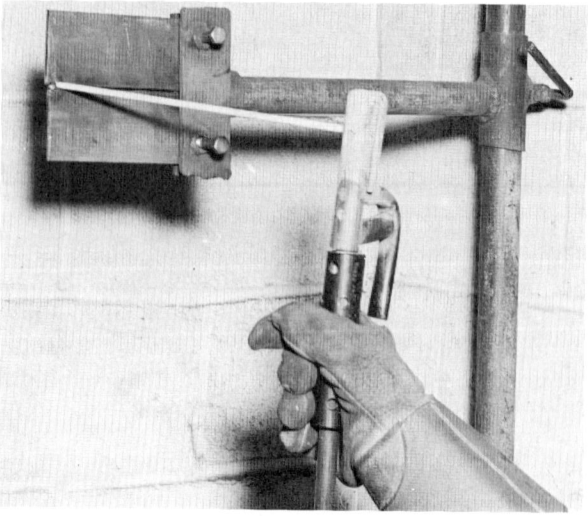

To perform welds of this kind, you must use a slightly shorter arc at a slight reduction in amperage setting than you would for flat position welding. The shorter arc will minimize the tendency of the molten puddle to sag and cause overlaps. An overlap occurs when the puddle runs down to the lower side of the bead and solidifies on the surface without actually penetrating the metal. See Fig. 8-2. A sagging puddle usually leaves an undercut on the top side of the seam as well as improperly shaped beads, all of which weaken a weld.

Fig. 8-1. Welding a horizontal seam.

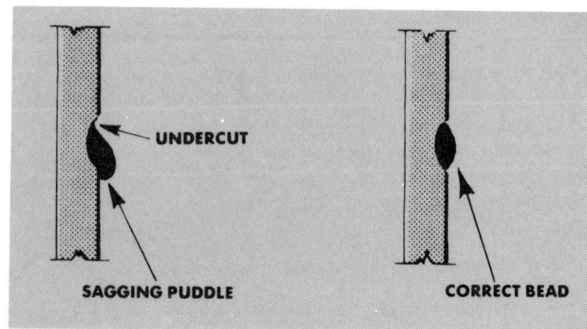

Fig. 8-2. The sagging puddle at the left can be avoided by shortening the arc.

Shielded Metal Arc—Horizontal Welding

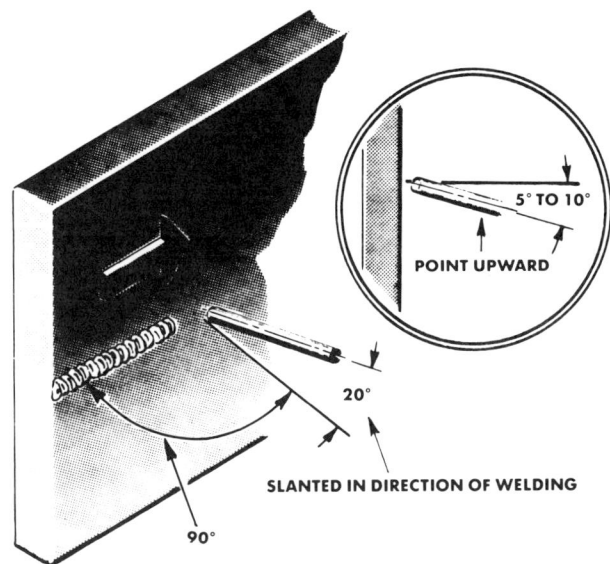

Fig. 8-3. Position of the electrode for horizontal welding.

How to hold the electrode. For horizontal welding, hold the electrode so that it points upward 5° to 10° and slants approximately 20° away from the deposited bead, as illustrated in Fig. 8-3. In laying the bead, use a narrow weaving motion as shown in Fig. 8-4. By weaving the electrode, the heat will be distributed more evenly, thereby reducing still further any tendency for the puddle to sag. Keep the arc as short as possible. If the force of the arc has a tendency to undercut the plate at the top of the bead, drop the electrode a little to increase the upward angle.

As the electrode is moved in and out of the crater, pause slightly each time it is returned. This keeps the crater small and the bead is less likely to drop.

Multiple-pass horizontal welds. Run a bead along the root of the joint without any weaving motion. Clean the slag and deposit a second bead, using a slight weaving motion and penetrating the first bead and the plate. See Fig. 8-5.

Clean the slag off the second bead and deposit a third layer. Notice that the third bead penetrates into the first and second layers as well as into the upright plate. This penetration is important; otherwise a weak weld will result.

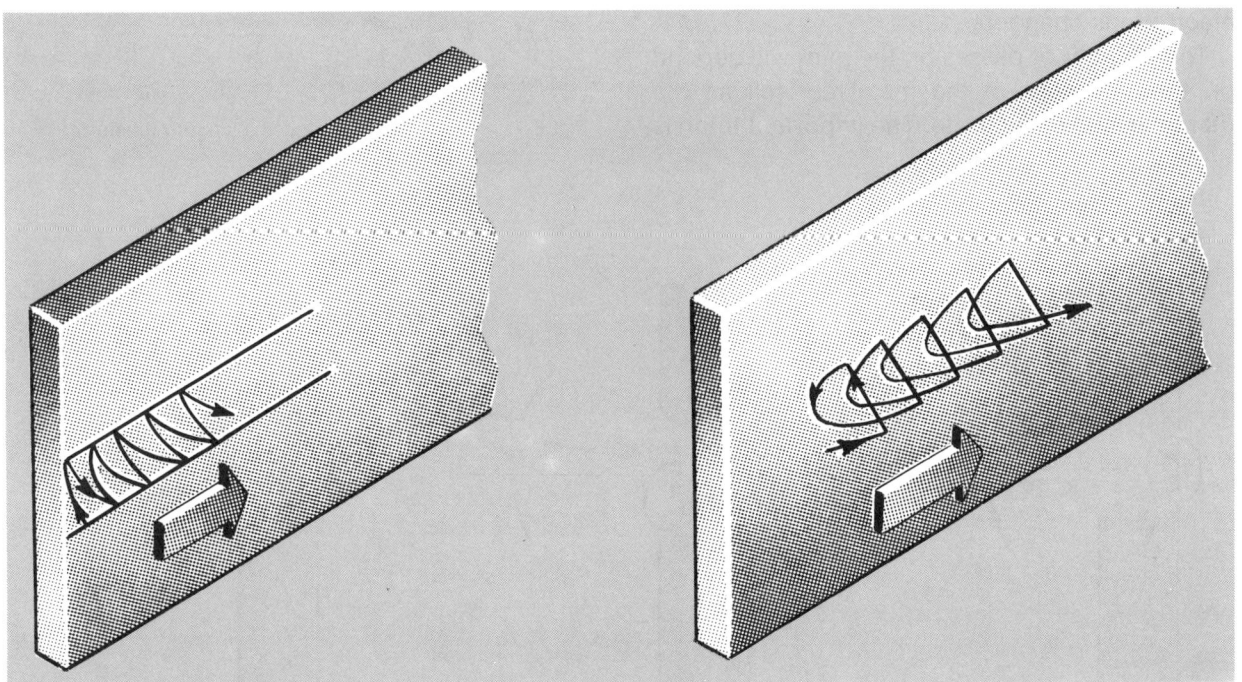

Fig. 8-4. The weaving pattern at left will result in a normal bead width, while the pattern at the right will result in a wider bead width.

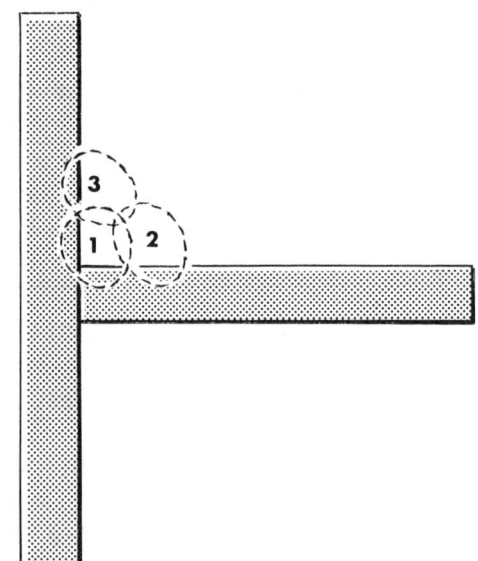

Fig. 8-5. Depositing a multiple bead on a T-joint.

to secure sufficient penetration into each adjacent layer. It is common practice on a wide joint to finish the weld with a *wash bead*, as illustrated in Fig. 8-7, to produce a smooth finish. A wash bead is made by using a wide weaving motion that covers the entire area of the deposited beads.

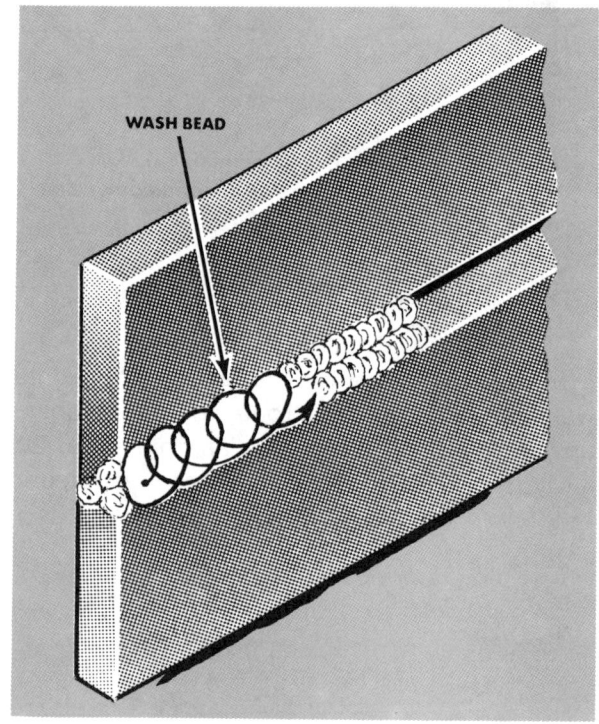

Fig. 8-7. A wash bead produces a smooth finish.

On many welding jobs, the practice is to bevel both edges of the joint to form a 60° included angle. Since such a joint does not provide a retaining shelf for the bead as the one shown in Fig. 8-5, a little more skill is required to produce a satisfactory weld. In practicing this type of weld, notice in Fig. 8-6 how the position of the electrode is changed.

The number of passes on the joint will depend on the thickness of the metal as well as the diameter of the electrode. The important thing is

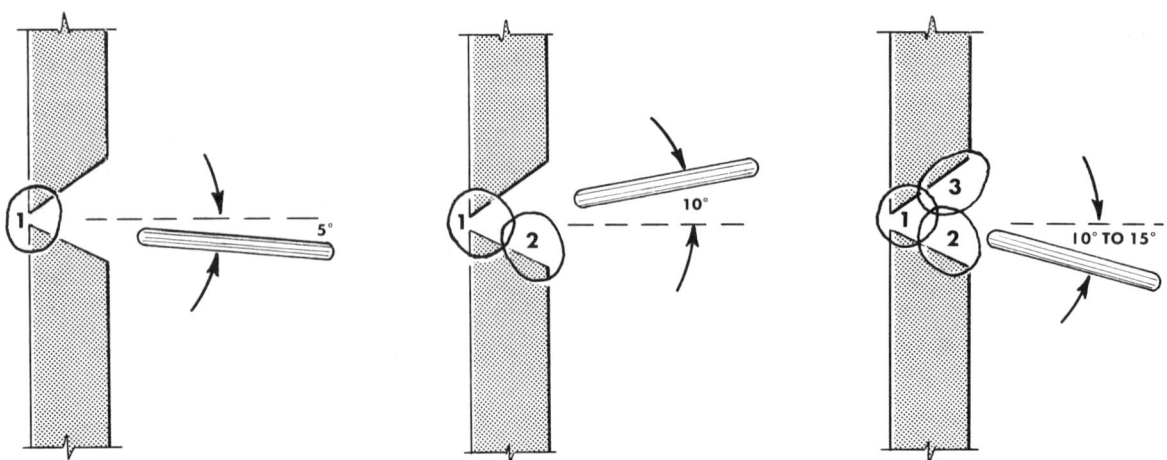

Fig. 8-6. Position of the electrode in horizontal butt welding.

SELF QUIZ

Correct answers are listed in the back of the book.

True and False

Circle the letter T if the statement is true or F if it is false.

1. T F A short arc is necessary in making a horizontal weld.
2. T F A sagging puddle usually leaves an undercut on the top side of a weld.
3. T F Overlaps on horizontally made welds will not particularly affect the quality of a weld.
4. T F In horizontal welding the electrode is held at about the same position as in flat position welding.
5. T F The amperage setting should be reduced slightly for horizontal welding.
6. T F A fast freeze type electrode is often very useful for making sound welds in a horizontal position.
7. T F If the molten metal begins to sag when making a horizontal weld, push the electrode deeper into the puddle.
8. T F A slight crescent type weaving motion is often used in horizontal welding.
9. T F If the electrode is manipulated properly when welding in the horizontal position, the formation of overlaps is practically eliminated.
10. T F An undercut is not particularly significant in horizontal welding.

WELDING ASSIGNMENT

I. Running Straight Surface Beads in Horizontal Position

1. Draw a series of parallel lines on a metal plate. Clamp the plate in a jig so welding can be done on about eye level.

2. First practice depositing continuous beads without any weaving motion. Use several type electrodes, such as E-6010, E-7014, E-7018, E-7024 and E-7028. Then try weaving the electrode. Use both weaving patterns as shown in Fig. 8-8. Notice how much wider the bead is

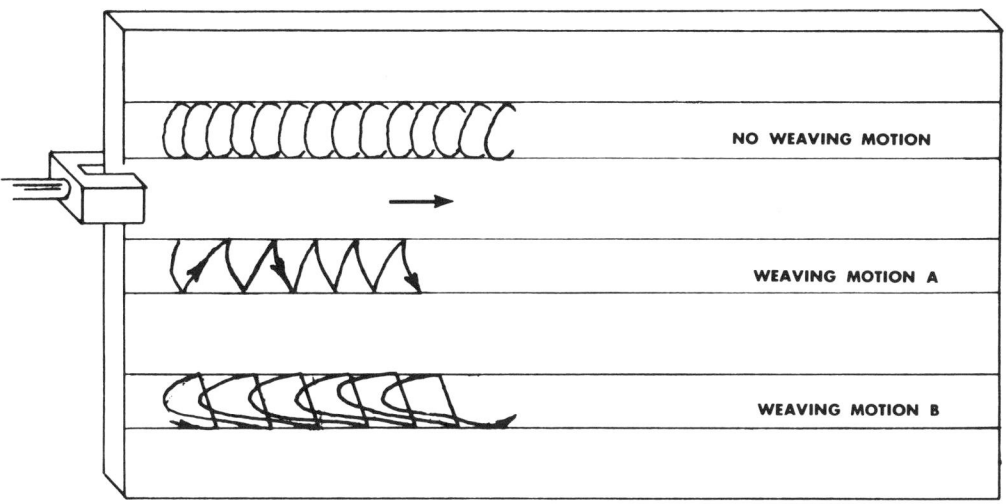

Fig. 8-8. Practice plate for running beads in a horizontal position.

when pattern B is used. Remember, keep the arc as short as possible. Experiment with the amperage switch to see which setting provides the best heat control.

3. Continue to practice this operation until you can produce uniform beads without overlaps and undercuts.

II. Making a Single-Pass Lap Joint in a Horizontal Position

1. Tack two plates to form a lap joint. Clamp the piece in a vertical position as shown in Fig. 8-9.

2. Run a single bead along the edge, using a slight weaving motion.

Watch the surface of the top plate closely to prevent any undercutting. Continue this operation on additional lap joints until a satisfactory weld is made.

3. Test each weld and analyze the results by referring to the Weld Analysis Check List shown at the end of this unit.

III. Welding a Single-Pass T and Butt Joint in a Horizontal Position

1. Use the same plate thicknesses and electrodes you used in the previous assignment.

2. First practice welding T joints and then butt joints. See Fig. 8-10. In each case confine the weld to a single pass.

3. Test each joint and analyze the results.

4. Continue to practice until your instructor thinks you have mastered this skill.

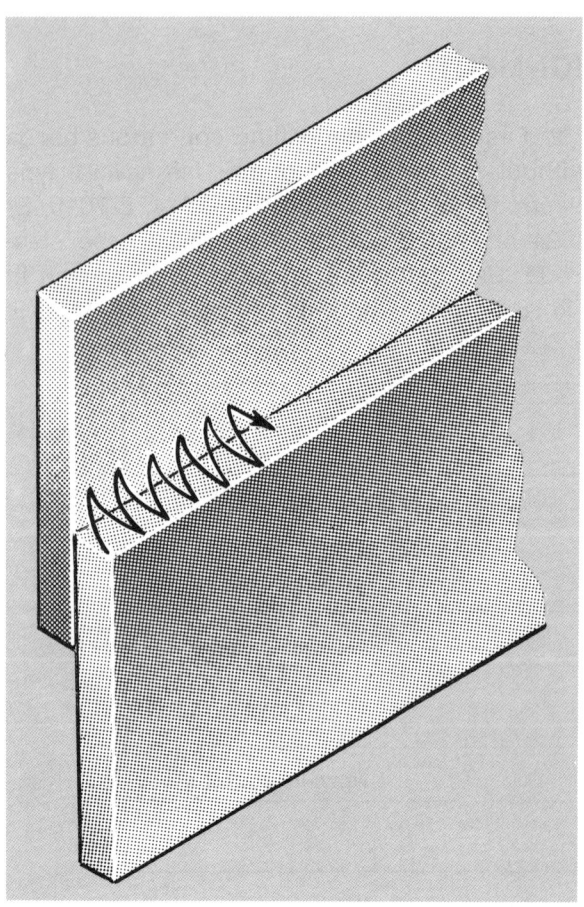

Fig. 8-9. Practice plate for a horizontal lap weld.

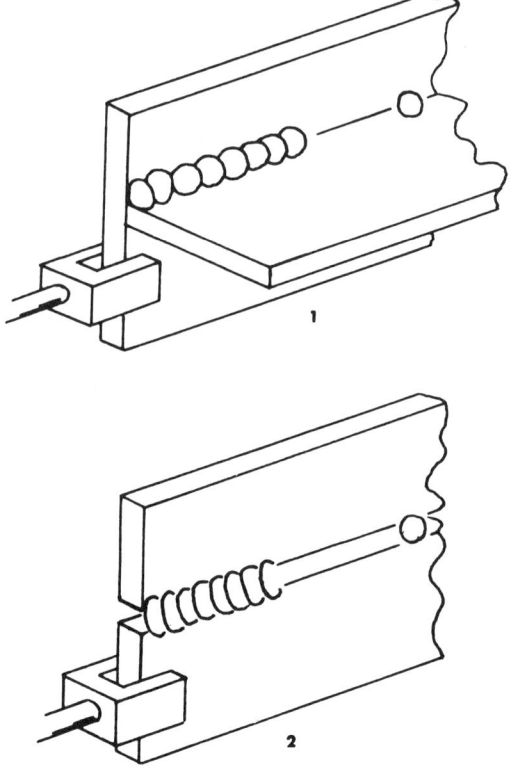

Fig. 8-10. Practice plates for welding horizontal joints.

IV. Welding a Multiple-Pass Butt Joint in a Horizontal Position

1. Obtain two pieces of ¼″ steel plate and *bevel the edge of one plate.*

2. Tack the two plates together to form a butt joint, allowing ¹⁄₁₆″ space at the root opening. Fasten the plates in a vertical position with the beveled plate on top as shown in Fig. 8-11. The plate that is not beveled should be on the bottom since its flat edge serves as a shelf, thus helping to prevent the molten metal from running out of the joint.

3. Deposit the first bead deeply in the root of the joint. Remove the slag and lay the second bead. Then follow with a third bead.

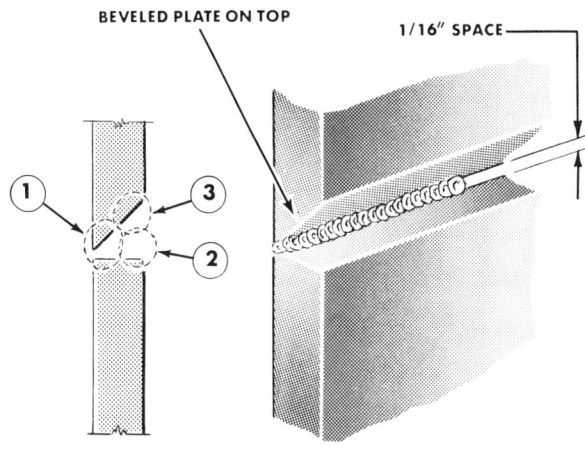

Fig. 8-11. Position of the butt joint for horizontal welding.

Weld Analysis Check List	Yes	No
1. Bead widths are right size		
2. Beads have uniform ripples		
3. Weld beads are too flat		
4. Weld beads are too high		
5. Weld penetration is insufficient		
6. Weld penetration is excessive		
7. Cold laps on surface		
8. Weld has surface porosity		
9. Weld has subsurface porosity		
10. Weld has crater cracks		
11. Weld has burn thru		
12. End crater is filled		
13. Weld passed bend test without cracking		

UNIT 9
Shielded Metal Arc—Vertical Welding

In the fabrication of many structures such as steel buildings, bridges, tanks, pipelines, ships, and machinery, the operator must frequently make vertical welds. A vertical weld is one with a seam or line of weld running up and down as shown in Fig. 9-1.

Fig. 9-1. After tacking the metal strips together, this operator lays vertical welds. (Hobart Brothers Co.)

One of the problems of vertical welding is that gravity tends to pull down the molten metal from the electrode and plates being welded. To prevent this from happening, fast-freeze types of electrodes should be used. Puddle control can also be achieved by proper electrode manipulation and selecting electrodes specifically designed for vertical position welding.

Electrode Position and Movement

Vertical welding is done by depositing beads either in an upward or downward direction (sometimes referred to as *uphill* and *downhill*). *Downward welding* is very practical for welding light gage metal because penetration is shallow, thereby forming an adequate weld without burning through the metal. Moreover, downward welding can be performed much more rapidly, which is important in production work. Although it is generally recommended for welding lighter materials, it can be used for most metal thicknesses.

On heavy plates of 1/4" or more in thickness *upward welding* is often more practical, since deeper penetration can be obtained. Welding upward also makes it possible to create a shelf for successive layers of beads.

Downhill welding. For downhill welding, tip the electrode as in Fig. 9-2 left. Start at the top of the seam and move downward with little or no weaving motion. If a slight weave is necessary, swing the electrode so the crescent is at the top. See Fig. 9-3. Keep the arc short and drag the electrode downward to form the bead. Travel just fast enough to keep the molten metal and slag from running ahead of the crater. Do not use any weaving motion to start with. Once this technique is mastered, try weaving the electrode but very slightly with the crest at the top of the crater. See Fig. 9-3. Stringer beads or small weaves are always preferred to wide weave passes. Use a low current. Point the electrode directly into the joint.

Uphill welding. For uphill welding, start with the electrode at right angles to the plates. Then, lower the rear of the electrode, keeping the tip in place, until the electrode forms an angle of 10°–15° with the horizontal line, as shown in Fig. 9-2, right.

On many vertical seams in uphill welding, it is necessary to form beads of various widths. The width of the bead can be controlled by using one of the weaving patterns shown in Fig. 9-4. Each pattern will produce a bead approximately twice the diameter of the electrode. Notice that each weave is shaped so the electrode can dig into the metal at the bottom of the stroke, and the upward motion momentarily removes the heat until

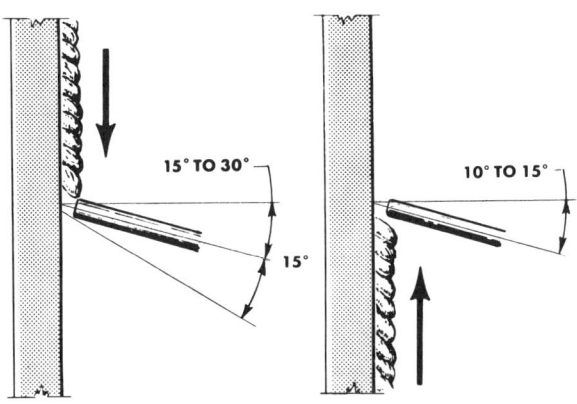

Fig. 9-2. Position of the electrode for downward (left) and upward (right) vertical welding.

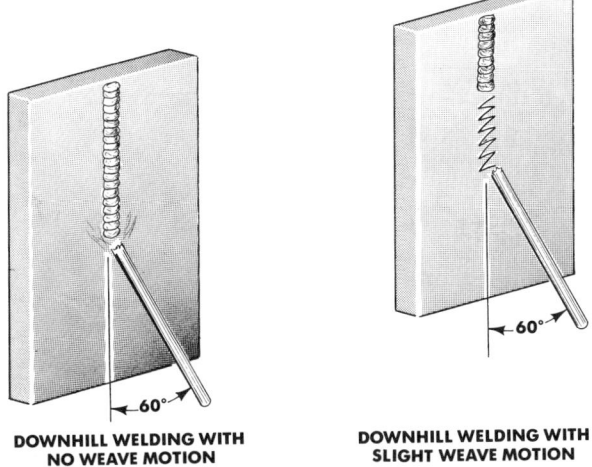

Fig. 9-3. Downhill welding methods.

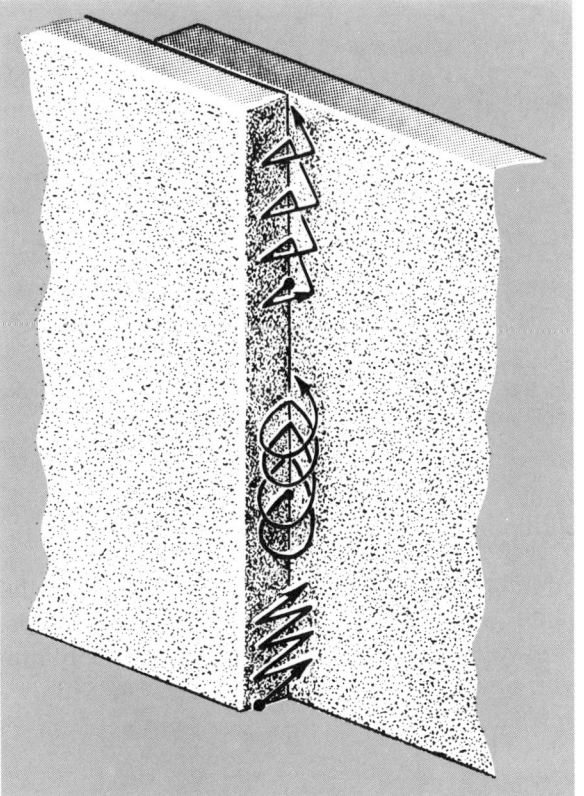

Fig. 9-4. Weaving patterns can be used to vary bead width.

80 Arc Welding

the metal can solidify. When a smooth weld is required on the final pass of a wide joint, the *wash bead* should be used.

Another method used by welders to control the temperature of the molten puddle is a *whipping* action of the electrode. This technique is especially helpful when welding pieces that do not have a tight fit and large openings have to be filled. Also, it is used in overhead and vertical welding to control the weld puddle better.

In a whipping action the electrode is struck and held momentarily. Then it is moved forward about $1/4''$ or $3/8''$ and raised a similar distance at the same time. Raising the electrode temporarily reduces the heat. Then just as the puddle begins to freeze, the electrode is moved back into the center of the puddle and the sequences are repeated. The movement of the electrode should be done by pivoting the wrist and not moving the arm while making the pass. See Fig. 9-5.

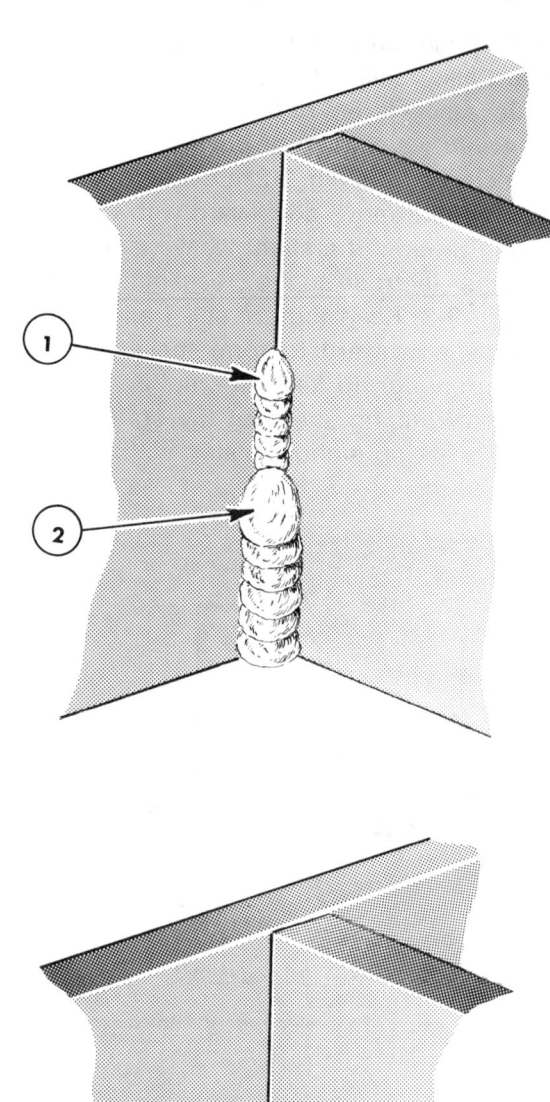

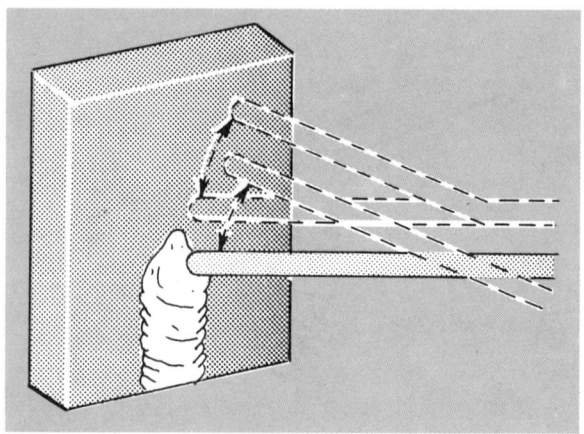

Fig. 9-5. A whipping motion helps to control the puddle in uphill welding.

Multiple Pass Vertical Welds

Multiple-pass vertical welds are made in the same way as multiple-pass horizontal welds. First you make a stringer bead and you continue with one or more filler passes. See Fig. 9-6.

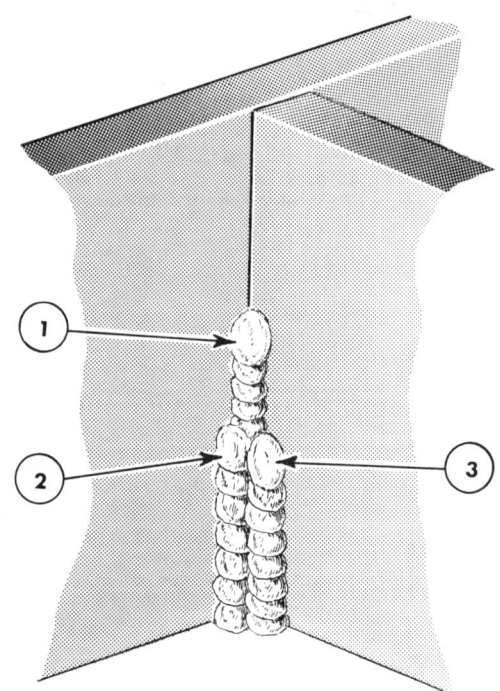

Fig. 9-6. Two and three-pass vertical welds.

SELF QUIZ

Correct answers are listed in the back of the book.

Short Answer

In the blanks provided write the answers to the following questions. Use as few words as possible.

1. What is the main thing that must be avoided in vertical welding? _____

2. Why is downward welding more practical for welding light gage materials? _____

3. How much should the electrode be tipped downward for vertical welding? _____

4. Generally, should any weaving motion be used in downward welding? _____

5. If a wider bead is necessary when making a downward weld, how can this be achieved? _____

6. Why is the upward welding technique more practical on heavy plates? _____

7. How is the electrode often manipulated when welding upward? _____

8. Regardless of the weaving pattern used in upward welding, why is the electrode swung slightly up and away from the puddle on the top of its stroke? _____

9. A weaving motion will usually produce a bead of approximately what width? _____

10. What type electrodes are most practical for downhill and uphill welding? _____

WELDING ASSIGNMENT

I. Depositing Surface Beads in Vertical Position

1. Lay out several guide lines on a suitable metal plate for depositing continuous vertical beads. Fasten the plate in a jig for vertical welding. See Fig. 9-7.

2. First practice running straight beads from top to bottom without any weaving motion. On some occasions a slight crescent motion is used in downward welding to make a wider bead. Practice this technique as well. Since downhill welding is best for joining thin materials, the use of fill-freeze or fast-freeze electrodes, such as E-6010, E-6012, E-6013 or E-7014, will minimize burn thru.

3. After the downward method is mastered, practice laying continuous beads from bottom to the top. Start without using any weaving motion. Then practice the rocking motion and other weaving patterns as described in this book.

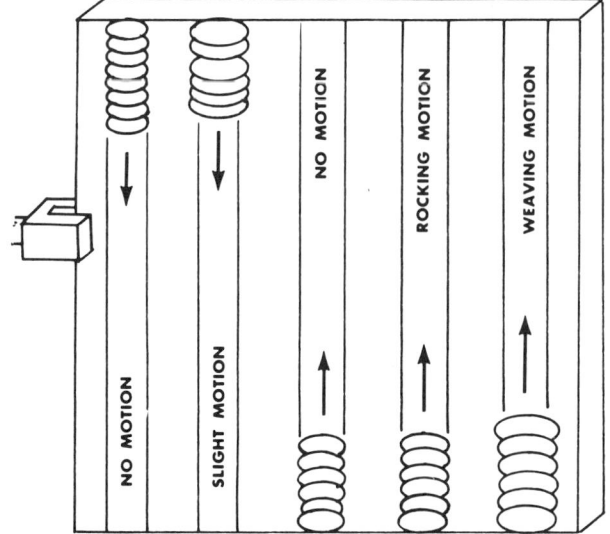

Fig. 9-7. Practice plate for depositing vertical surface beads.

82 Arc Welding

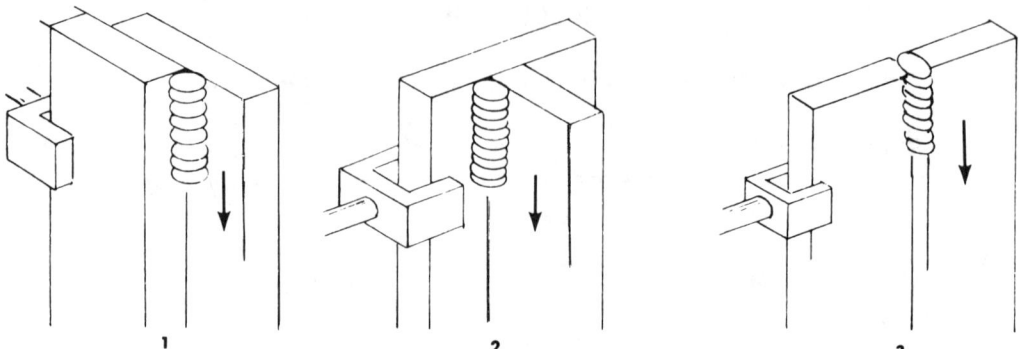

Fig. 9-8. Practice plates for downward welding.

II. Welding Vertical Joints—Downward Method, Single Pass

1. Select several metal plates 1/8" or less in thickness. Tack weld them to form a lap joint, a T joint, and an open-butt joint. See Fig. 9-8.

2. Start welding a lap joint first and continue with a T joint and finally a butt joint. Use a single pass on all joints. On the first set of joints do not use any weaving motion. On the second set practice a slight crescent motion. Weld on one side only.

3. Test each weld and evaluate the results using the Weld Analysis Check List shown at the end of this unit.

4. Continue welding these joints with a single pass until they are approved by your instructor.

III. Welding Vertical Joints—Upward Method, Single Pass

1. Practice welding lap, T, and open-butt joints using the upward method. See Fig. 9-9.

2. Use a single pass on each joint, first without any weaving motion and then with a rocking motion. Weld on one side only.

3. Test each welded joint and evaluate the results.

4. Continue to practice making these welds until they are certified by your instructor.

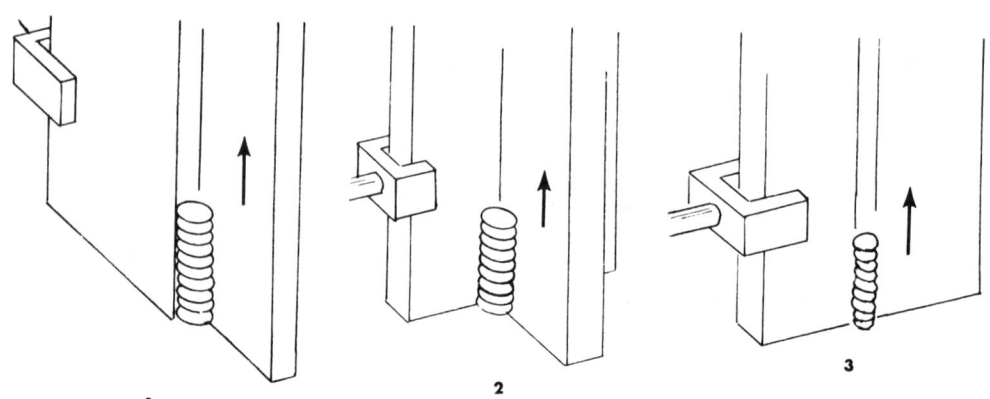

Fig. 9-9. Practice plates for upward welding.

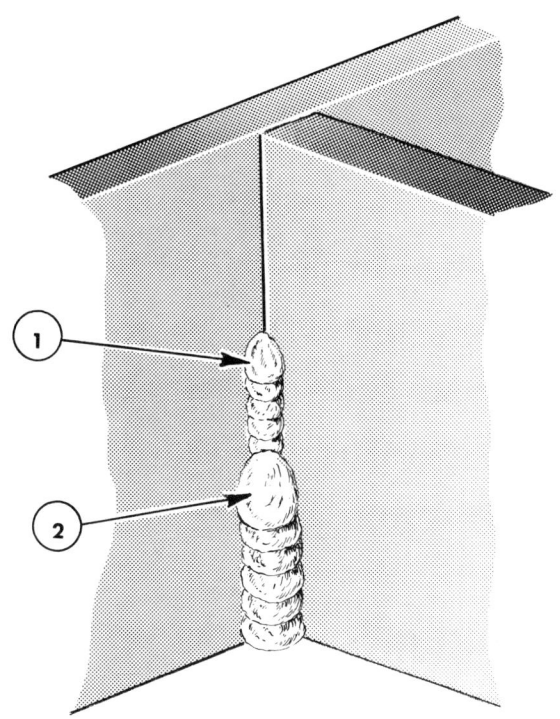
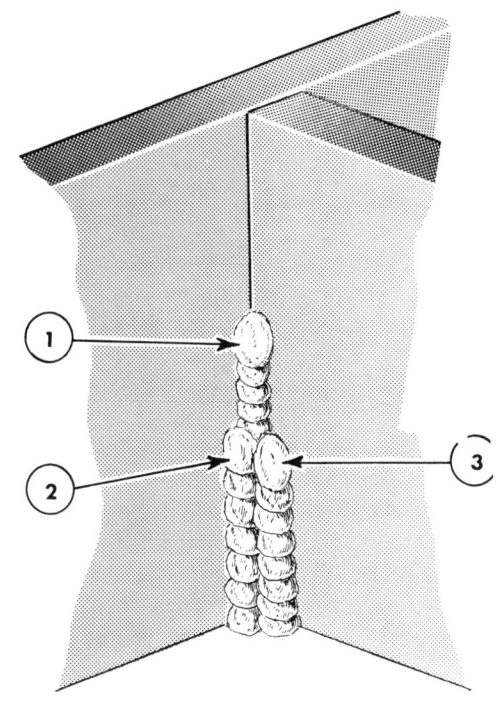

Fig. 9-10. Practice plate for a multiple vertical weld.

IV. Making a Multiple-Pass Vertical Weld

1. Obtain two 1/4" plates and tack them to form a T-joint. Support the joint in a vertical position.

2. Deposit a narrow, straight bead in the root.

3. Remove the slag and deposit at least one or two additional layers as shown in Fig. 9-10.

Weld Analysis Check List	Yes	No
1. Bead widths are right size		
2. Beads have uniform ripples		
3. Weld beads are too flat		
4. Weld beads are too high		
5. Weld penetration is insufficient		
6. Weld penetration is excessive		
7. Cold laps on surface		
8. Weld has surface porosity		
9. Weld has subsurface porosity		
10. Weld has crater cracks		
11. Weld has burn thru		
12. End crater is filled		
13. Weld passed bend test without cracking		

UNIT 10
Shielded Metal Arc—Overhead Welding

Welding in an overhead position is probably the most difficult operation to master. It is difficult because you must assume an awkward stance and at the same time work against gravity, which exerts a downward force. See Fig. 10-1. In an overhead position the puddle has a tendency to drop, making it harder to secure uniform beads and correct penetration. Nevertheless, with a little practice it is possible to secure welds as good as those made in other positions.

Fig. 10-1. Overhead welding requires a little more skill than vertical position welding.

Shielded Metal Arc—Overhead Welding

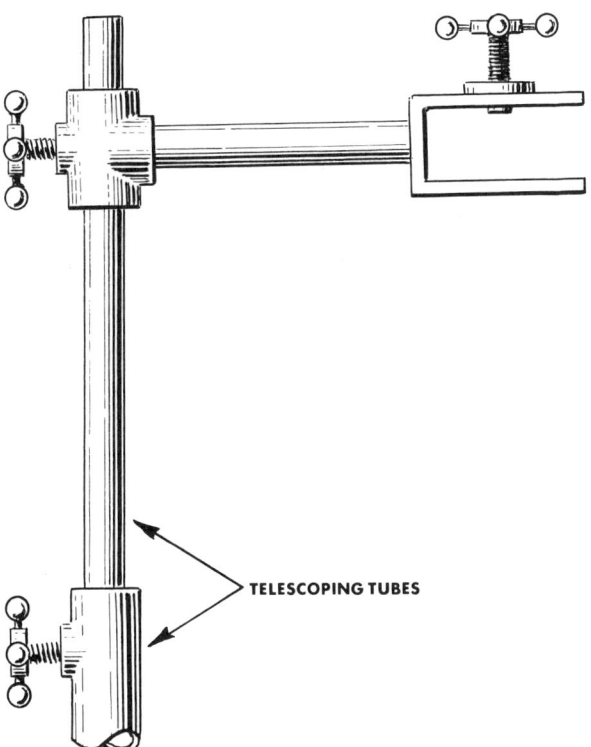

Fig. 10-2. A jig is necessary to raise the work to the correct height and position.

In learning to weld in an overhead position, some kind of fixture will be needed. The one shown in Fig. 10-2 is recommended, since the work can be adjusted to any height or position. *Caution: Since there is a possibility of some falling molten metal, one's clothing should be checked very carefully. Be sure the sleeves and pant cuffs are rolled down and a protective garment with a tight-fitting collar is zipped or buttoned up to the neck. Also, it is a good idea to wear a cap and heavy duty shoes.*

Position for Overhead Welding

To start welding, place the electrode in the holder and hold it at right angles to the seam. Then tilt the rear of the electrode away from the crater until the electrode forms an angle of 10° to 15° with the horizontal line as shown in Fig. 10-3. The line of weld may be in any direction—forward, backward, left, or right.

Grip the holder so the knuckles are up and the palm down. This prevents particles of molten metal from being caught in the hollow palm of

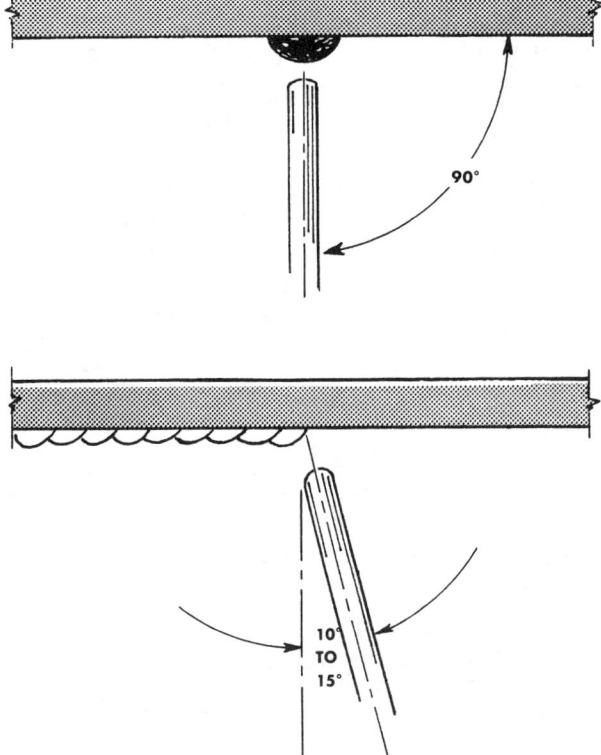

Fig. 10-3. Position of the electrode for overhead welding.

86 Arc Welding

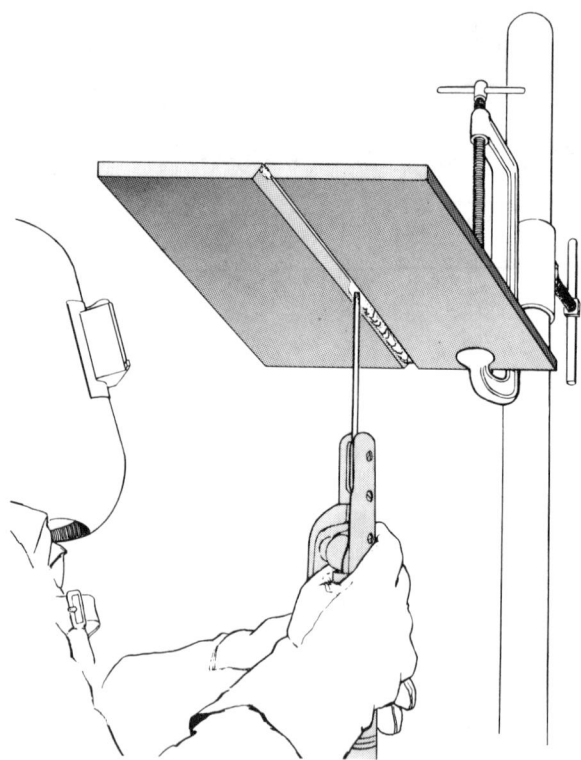

Fig. 10-4. Sometimes it is better to use both hands in overhead welding.

the glove and allows the spatter to roll off the glove. Although the electrode can be held in one hand, sometimes it is better if it is held with both hands. See Fig. 10-4. To gain as much protection as possible from falling sparks and hot metal drippings, stand to the side rather than directly underneath the arc. The discomfort of the cable

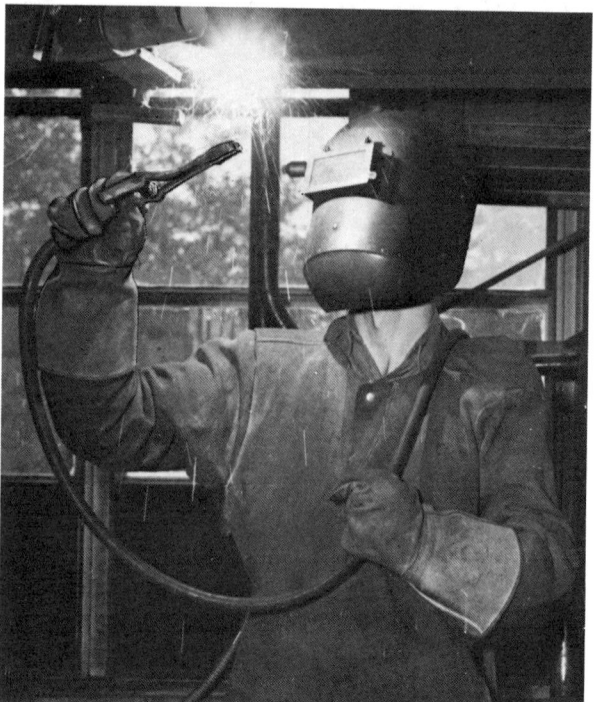

Fig. 10-5. If you are standing, drape the cable over your shoulder. (Hobart Brothers Co.)

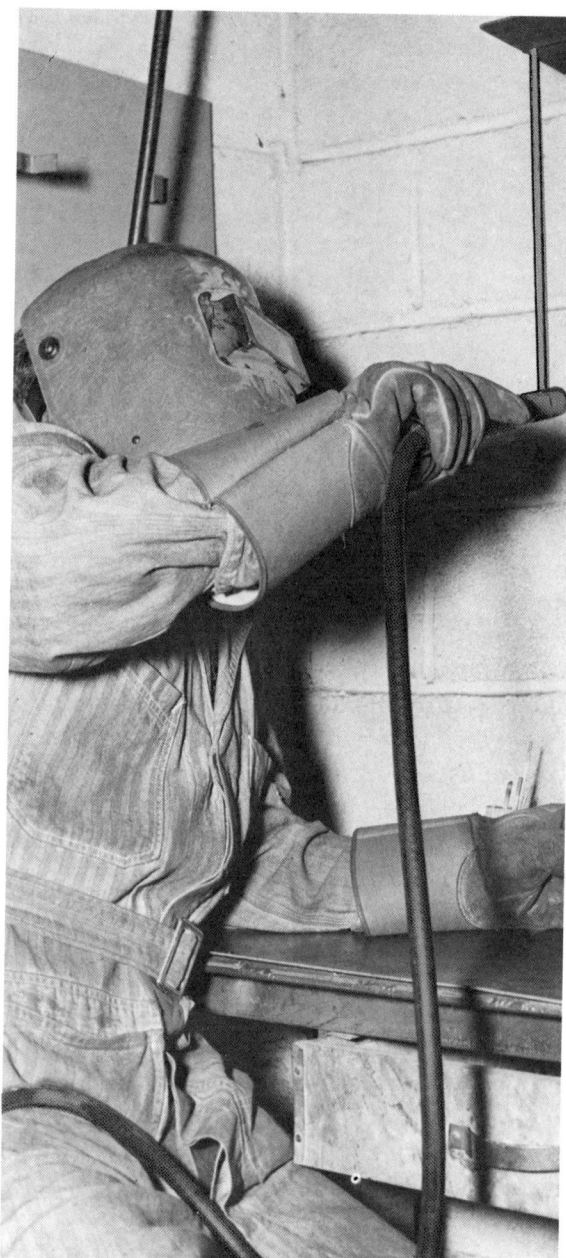

Fig. 10-6. If you are sitting, drape the cable over your knees.

can be minimized by draping it over the shoulder, if you are welding in a standing position, or over the knees if in a sitting position. See Figs. 10-5 and 10-6.

Welding procedure. Welding in the overhead position is actually no different from welding in any other position. In the overhead and horizontal positions, however, the welder has to keep the puddle from sagging. The use of fast-freeze electrodes and lower current will help control the puddle. The two common motions used in overhead welding are shown in Fig. 10-7.

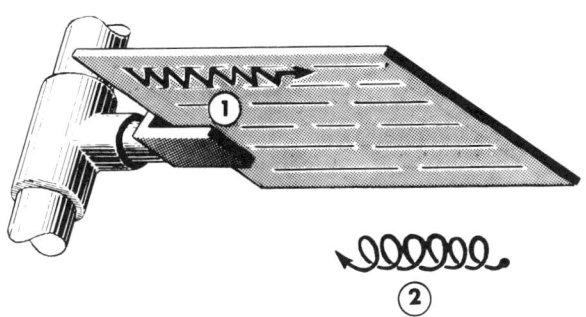

Fig. 10-7. Two kinds of weaving motions are used in overhead welding.

SELF QUIZ

Correct answers are listed in the back of the book.

Short Answer

In the blanks provided write the answers to the following questions. Use as few words as possible.

1. What kind of electrode is best for overhead welding? _____
2. In what direction should the palm of your hand that holds the electrode holder be facing when welding overhead? _____
3. Is it advisable to hold the electrode holder with both hands on some overhead welding jobs? _____
4. How should the cable hang when welding overhead while in a standing position? _____
5. In addition to keeping a short arc, what else can be done to keep the puddle from dropping in overhead welding? _____
6. What is the largest diameter electrode that should be used in overhead welding? _____
7. How much should the electrode be inclined away from the crater in overhead welding? _____
8. Why should the amperage be set at the lower end of its range for the electrode involved? _____
9. How should you position yourself when making an overhead weld? _____
10. Why is proper clothing particularly important in overhead welding? _____

WELDING ASSIGNMENT

I. Running Overhead Beads

1. Lay out a series of guide lines on a metal plate and fasten the plate in a suitable jig.

2. Deposit continuous beads over the guide lines. Experiment with the amperage setting to get the best possible heat control.

3. First practice laying beads without any

88 Arc Welding

electrode motion. Then try to use a rocking motion. Continue by depositing beads and using the two weaving patterns shown in Fig. 10-8.

4. Continue laying straight beads until you are able to control the puddle and make uniform beads.

3. Lay the first bead deep in the root of the joint.

4. Remove the slag and deposit the second bead on the wide side of the plate. Clean off the slag and deposit the third bead. See Fig. 10-9.

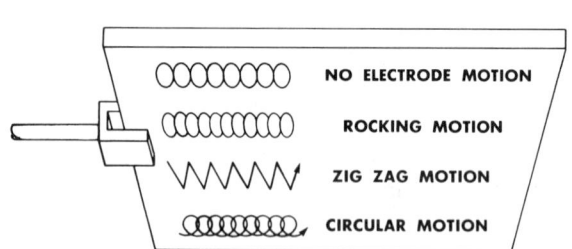

Fig. 10-8. Practice plate for running overhead beads.

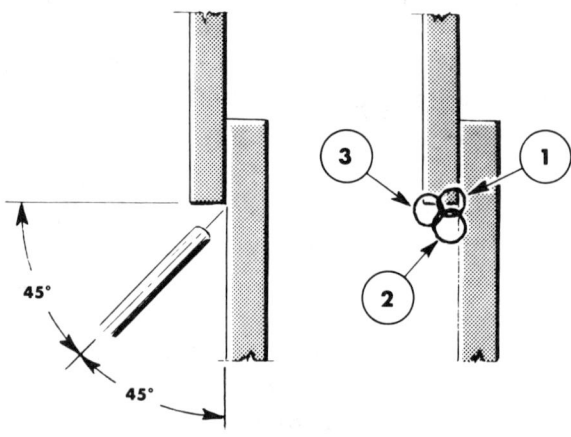

Fig. 10-9. Practice plate for multiple-pass lap joint.

II. Welding Joints in an Overhead Position—Single Pass

1. Tack weld available metal plates to form lap, T, and butt joints.

2. Start welding a lap joint first, then a T joint, and finally a butt joint. Use a single pass on each joint.

3. Test each joint and evaluate the results using the Weld Analysis Check List shown at the end of this unit.

4. Continue to practice welding these joints in the overhead position until your instructor certifies that you have acquired sufficient skill.

III. Welding a Lap Joint in an Overhead Position—Multiple Pass

1. Tack two 1/4" plates to form a lap joint and clamp them in the overhead jig.

2. Hold the electrode so it bisects the angle between the plates and is inclined slightly away from the crater as illustrated in Fig. 10-9.

IV. Welding a T Joint in an Overhead Position—Multiple Pass

1. Tack two 1/4" plates to form a T joint and clamp them in the overhead jig.

2. Deposit the first bead in the root of the V. Clean off the slag and deposit two additional beads as shown in Fig. 10-10.

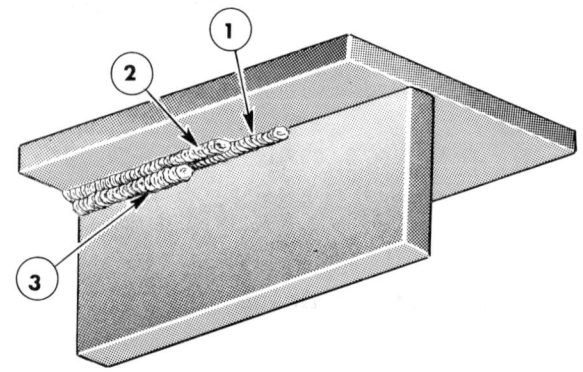

Fig. 10-10. Practice plate for multiple-pass T-joint.

Weld Analysis Check List

	Yes	No
1. Bead widths are right size	___	___
2. Beads have uniform ripples	___	___
3. Weld beads are too flat	___	___
4. Weld beads are too high	___	___
5. Weld penetration is insufficient	___	___
6. Weld penetration is excessive	___	___
7. Cold laps on surface	___	___
8. Weld has surface porosity	___	___
9. Weld has subsurface porosity	___	___
10. Weld has crater cracks	___	___
11. Weld has burn thru	___	___
12. End crater is filled	___	___
13. Weld passed bend test without cracking	___	___

UNIT 11
Gas Tungsten-Arc Welding—GTAW

The gas tungsten-arc, commonly known as Tig, is used extensively in welding light gage materials. The process consists of an arc that is generated by means of a non-consumable tungsten electrode and a gas which provides a shield over the weld to protect it from atmospheric contamination. The weld can be made with or without a filler rod. See Fig. 11-1.

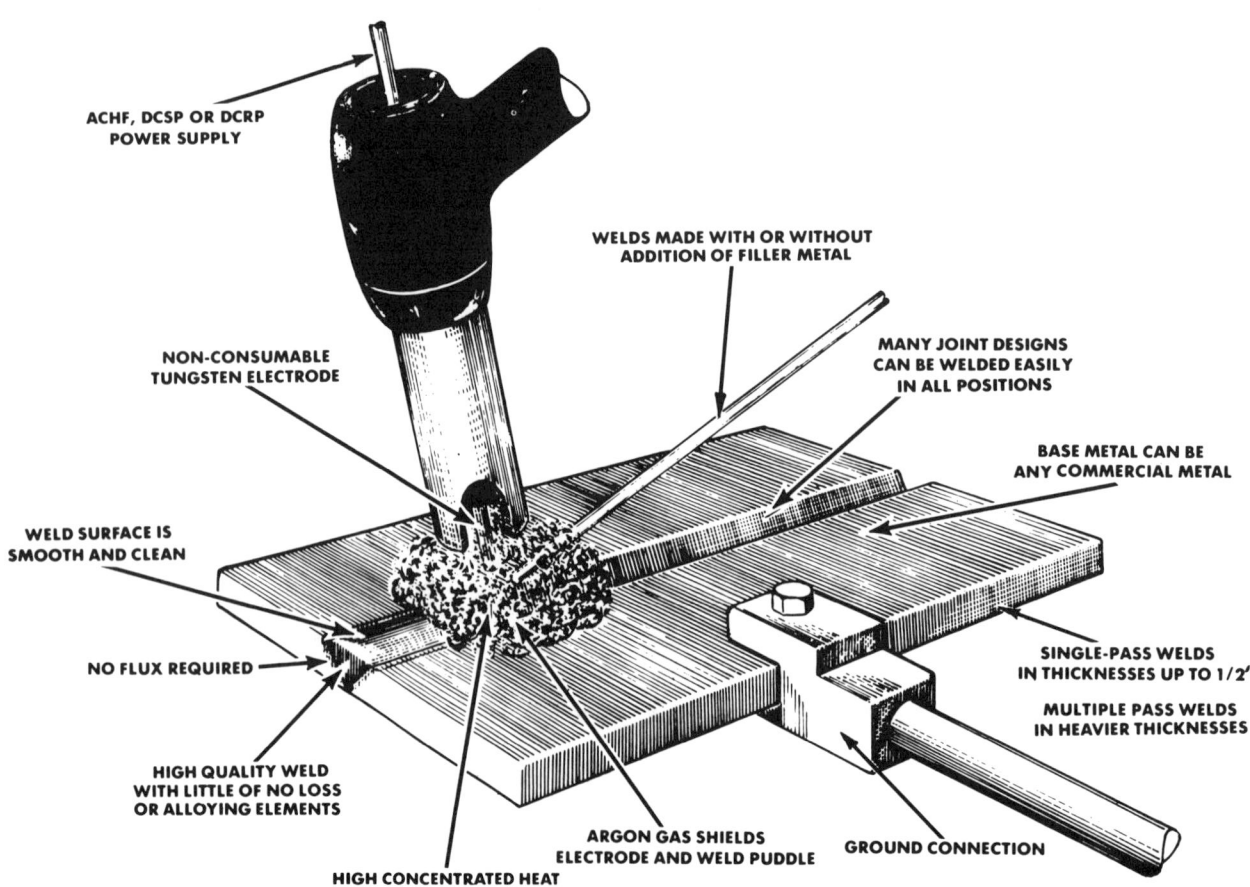

Fig. 11-1. In Tig welding, a non-consumable tungsten electrode is used. It is surrounded by a shield of inert gas. (Linde Co.)

Originally Tig was developed to join difficult-to-weld metals, such as aluminum, magnesium, stainless steels, copper and others. Today it is a standard process in welding all widely used commercial metals. Its greatest advantage is that welds can be made easily without fear of slag inclusions. Unlike the shielded metal-arc where slag forms over the weld, no such covering forms over Tig-made welds. Furthermore, Tig welds are usually stronger, because there is less danger of oxygen and nitrogen entering the molten puddle to cause porosity. Tig welding can be carried out in any position.

In Tig welding, the electrode is used only to create the arc. It is not consumed in the weld. In this way it differs from the regular shielded metal-arc process, where the stick electrode is consumed in the weld. For joints where additional weld metal is needed, a filler rod is fed into the puddle in a manner similar to welding with the oxy-acetylene flame process.

This type of welding is often referred to as *Heliarc* (Linde) or *Heliwelding* (Airco), both of which are manufacturer's trade names.

Welding Machines

Specially designed machines with all the necessary controls are available for tungsten-arc welding. Many power supply units are made to produce both AC and DC current. See Fig. 11-2.

The choice of an AC or DC machine depends on the weld characteristics desired. Some metals are joined more easily with AC current while others with DC current. See Table 11-1. To understand the effects of the two different currents, an explanation of their behavior in a welding process is necessary.

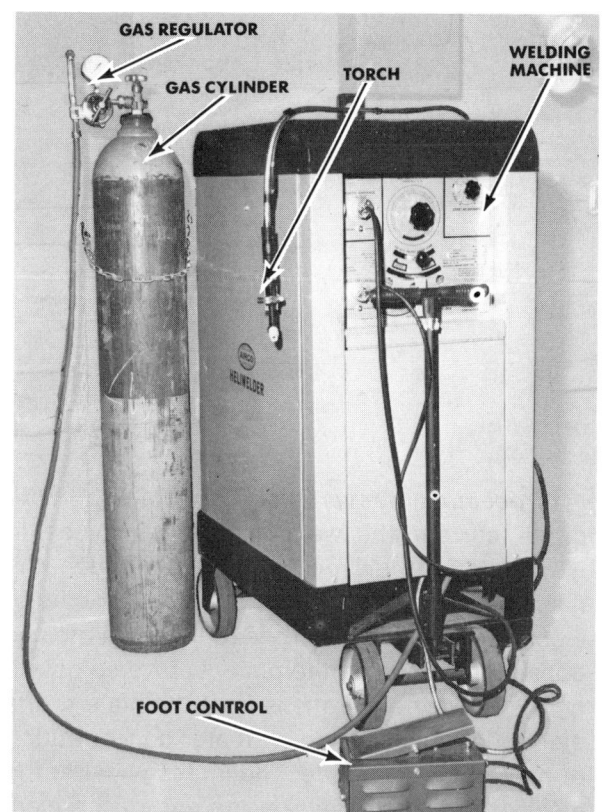

Fig. 11-2. A complete facility for tungsten inert gas (Tig) welding. (Miller Electric Manufacturing Co.)

TABLE 11-1. CURRENT SELECTION FOR TIG WELDING.

METAL	AC CURRENT with High Frequency Stabilization	DC CURRENT Straight Polarity	DC CURRENT Reverse Polarity
Magnesium up to 1/8" thick	1	NR	2
Magnesium above 3/16" thick	1	NR	NR
Magnesium castings	1	NR	2
Aluminum	1	NR	2
Aluminum castings	1	NR	NR
Stainless steel up to 0.050"	1	2	NR
Stainless steel 0.050" and up	2	1	NR
Brass alloys	2	1	NR
Silver	2	1	NR
Hastelloy alloys	2	1	NR
Silver cladding	1	NR	NR
Hard-facing	1	2	NR
Cast iron	2	1	NR
Low carbon steel 0.015" to 0.030"	2	1	NR
Low carbon steel 0.030" to 0.125"	NR	1	NR
High carbon steel 0.015" to 0.030"	2	1	NR
High carbon steel 0.030" and up	2	1	NR
Deoxidized copper up to 0.090"	NR	1	NR

Key: 1. Excellent operation—best recommendation
2. Good operation—second recommendation
NR = Not recommended

Direct current reverse polarity (DCRP). With direct current the welding circuit may be in either straight or reverse polarity. When the machine is set for straight polarity, the flow of electrons from the electrode to the plate creates considerable heat in the plate. In reverse polarity, the flow of electrons is from the plate to the electrode, thus causing a greater concentration of heat at the electrode. The intense heat at the electrode tends to melt off the end of the electrode and may contaminate the weld. Hence, for any given current, DCRP requires a larger diameter electrode than DCSP. For example, a 1/16" diameter tungsten electrode normally can handle about 125 amperes in a straight-polarity circuit. However, if reverse polarity is used with this amount of current, the tip of the electrode will melt off. Consequently, a 1/4" diameter electrode will be required to handle 125 amperes of welding current.

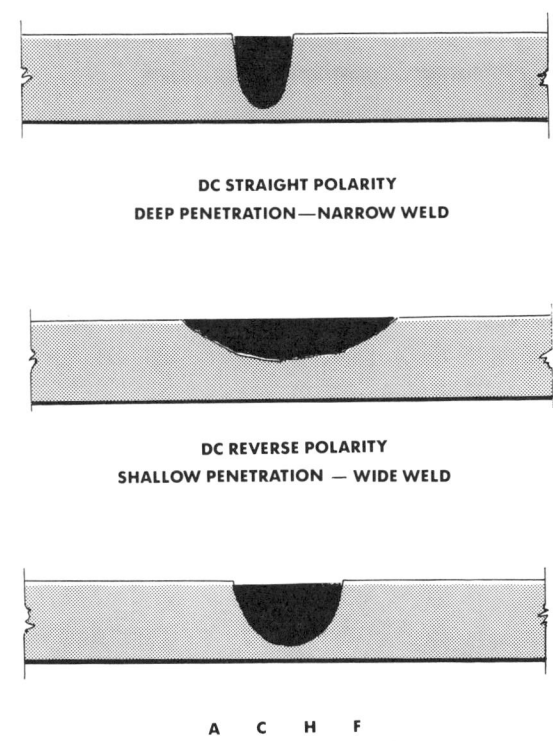

Fig. 11-3. The different types of operating current directly affect weld penetration, contour, and metal transfer. (Linde Co.)

Polarity also affects the shape of the weld. DCSP produces a narrow, deep weld, whereas DCRP with its larger diameter electrode and lower current forms a wide and shallow weld. See Fig. 11-3. For this reason *DCRP is never used in gas tungsten-arc welding except occasionally for welding aluminum and magnesium.* These metals have a heavy oxide coating which is more readily removed by the greater current-cleaning action of DCRP.

The same cleaning action is present in the reverse-polarity half of the AC welding cycle. No other metals require the kind of cleaning action that is normally needed on aluminum and magnesium. The cleaning action develops because of a bombardment of positive charged gas ions that are attracted to the negative charged workpiece. These gas ions when striking the metal have sufficient power to break the oxide and dislodge it from the surface. Generally speaking, better results are obtained in welding aluminum and magnesium with alternating current.

Direct current straight polarity (DCSP). Direct current straight polarity is used for welding most metals because better welds are achieved. With the heat concentrated at the plate, the welding process is more rapid, there is less distortion of the base metal, and the weld puddle is deeper and narrower than with DCRP. Since more heat is liberated at the puddle, smaller diameter electrodes can be used.

Alternating current (ACHF). Oxides, scale and moisture on the workpiece often tend to prevent the full flow of current in the reverse-polarity direction. During a welding operation the partial or complete stoppage of current flow (rectification) would cause the arc to be unstable and sometimes even to go out. To prevent such rectification, AC welding machines incorporate a high-frequency current flow unit. The high-frequency current is able to jump the gap between the electrode and the workpiece, piercing the oxide film and forming a path for the welding current to follow.

Although ACHF can be used to weld most metals, it is particularly effective for welding metals having a heavy oxide coating. See Table 11-1.

Torches

Manually operated welding torches are designed to conduct both the welding current and the inert gas to the weld zone. These torches are either air or water-cooled. See Fig. 11-4. Air-cooled torches are designed for welding light gage materials where low current values are used. Water-cooled torches are recommended when the welding requires amperages over 200A. A circulating stream of water flows around the torch to keep it from overheating.

The tungsten electrode which supplies the welding current is held rigidly in the torch by means of a collet that screws into the body of the torch. A variety of collet sizes are available so different diameter electrodes can be used. Gas is fed to the weld zone through a nozzle which

94 Arc Welding

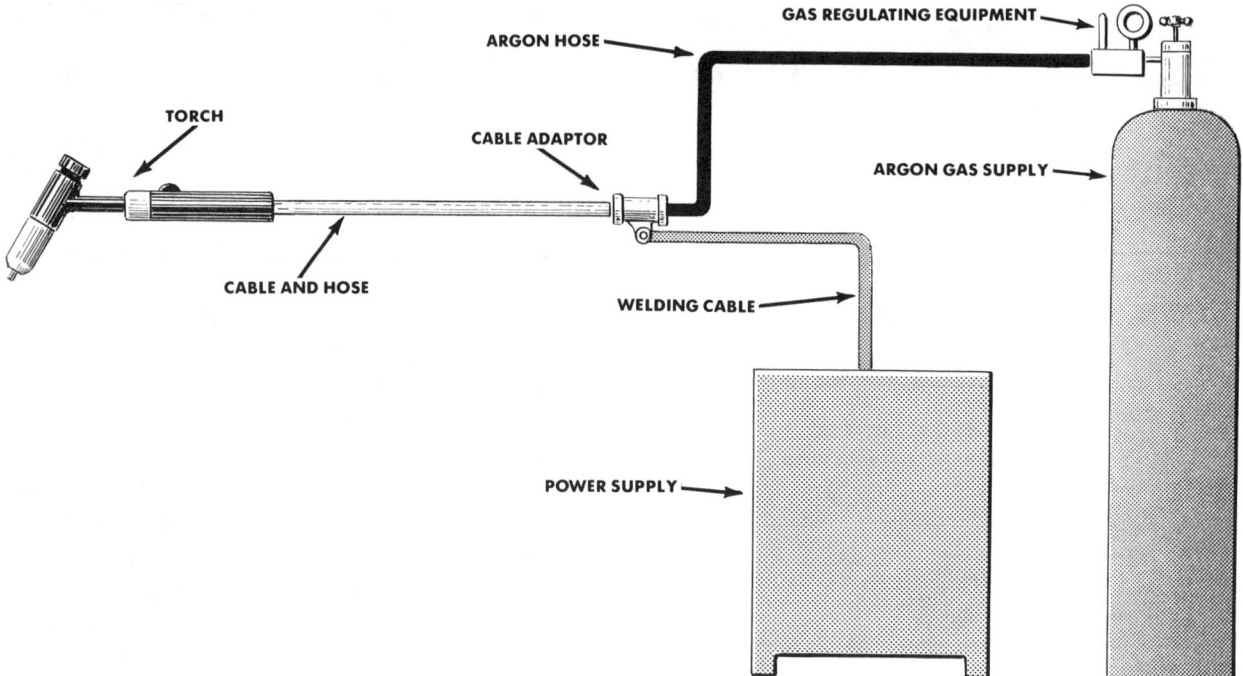

Fig. 11-4. A Tig welding unit with an air-cooled torch.

consists of a ceramic cup. Gas cups are threaded into the torch head to provide directional and distributional control of the shielding gas. The cups are interchangeable to accommodate a variety of gas flow rates.

Pressing a control switch on the torch starts the flow of both current and gas. On some equipment the flow of current and gas is energized by a foot control. The advantage of the foot control is that the current flow can be better controlled as the end of the weld is reached. By gradually decreasing the current it is less likely for a cavity to remain in the end of the weld puddle and less danger of cutting short the shielding gas.

Gas cups vary in size. The size to be used depends upon the type and size of the torch and the diameter of the electrode. See Table 11-2. (It is a good practice to follow the manufacturer's recommendations.)

Electrodes

Basic diameters of non-consumable electrodes are $1/16''$, $3/32''$, and $1/8''$. They are either pure tungsten, or alloyed tungsten. The alloyed tungsten electrodes usually have one to two percent thorium or zirconium. The addition of thorium increases the current capacity and electron emission, keeps the tip cooler at a given level of current, minimizes movement of the arc around the electrode tip, permits easier arc starting, and the electrode is not as easily contaminated by

TABLE 11-2. APPROXIMATE CUP ORIFICE FOR TIG WELDING.

TUNGSTEN ELECTRODE diameter (inches)	CUP ORIFICE diameter (inches)
1/16	1/4 –3/8
3/32	3/8 –7/16
1/8	7/16–1/2
3/16	1/2 –3/4

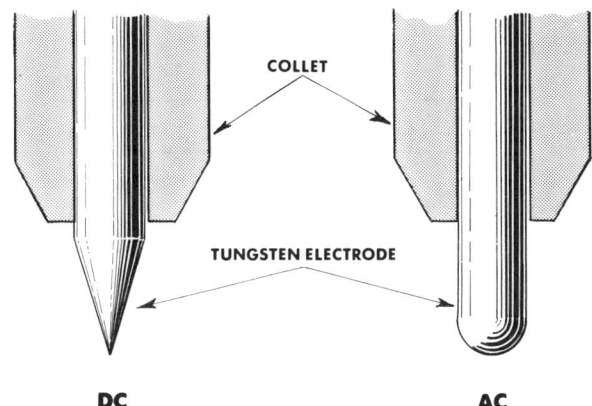

Fig. 11-5. Be sure the tungsten electrode is shaped with proper angle.

accidental contact with the workpiece. The two percent thoria electrodes normally maintain their formed point for a greater period than the one percent type. The higher thoria electrodes are used primarily for critical sheet metal weldments in aircraft and missile industries. They have little advantage over the lower thoria electrode for most steel welds. The introduction of the striped electrode combines the advantages of the pure, low, and high thoriated tungsten electrodes. This electrode has a solid stripe of two percent thoria inserted in a wedge along the full length of the electrode.

The diameter of the electrode selected for a welding operation is governed by the welding current to be used. Larger diameter tungsten electrodes are required with reversed polarity than with straight polarity. (See Tables 11-4, 11-5 and 11-6 for recommended sizes of electrodes, current, and material thickness for Tig welding.)

Electrode shapes. To produce good welds the tungsten electrode must be shaped correctly. The general practice is to use a pointed electrode with DC welding, and a spherical end with AC welding. See Fig. 11-5. It is also important that the electrode be straight; otherwise the gas flow will be off-center from the arc.

Shielding Gas

Shielding gas for gas tungsten-arc welding can be argon, helium, or a mixture of argon and helium. Argon is used more extensively because it is less expensive than helium. Argon is 1.4 times as heavy as air and 10 times as heavy as helium. There is very little difference between the viscosity of these two gases. Since argon is heavier than air, it provides a better blanket over the weld. Moreover, there is less clouding during the welding process with argon, and consequently the control of the weld puddle and arc is better.

Argon normally produces a better cleaning action, especially in welding aluminum and magnesium with alternating current. With argon there is a smoother and quieter arc action. The lower arc voltage characteristic of argon is particularly advantageous in welding thin material because there is less tendency to burn through the metal. Consequently, argon is used mostly for shielding purposes in welding materials up to $1/8$" in thickness.

The use of argon also permits better control of the arc in vertical and overhead welding. As a rule, the arc is easier to start in argon than in helium and for a given welding speed the weld produced is narrower with a smaller heat-affected zone. See Table 11-3 for the recommended selection of gases.

Argon and helium are supplied in steel cylinders containing approximately 330 cubit feet at a pressure of 2000 psi. A single or two-stage regulator may be used to control the gas flow. A

TABLE 11-3. SELECTION OF GASES.

METAL	TYPE	GAS	RESULT
Al	Manual welding	Argon	Better arc starting, cleaning action and weld quality; lower gas consumption
		Helium	High welding speeds possible
	Machine welding	Argon-Helium	Better weld quality, lower gas flow than required with straight helium
Mg	0–1/16″	Helium	Controlled penetration
	0–1/16″ +	Argon	Excellent cleaning, ease of puddle manipulation, low gas flows
Mild Steel	0–1/8″	Argon	Ease of manipulation, freedom from overheating
	0–1/8″ +		(Mig process preferred)
	Spot welding	Argon	Generally preferred for longer electrode life
			Better weld nugget contour
			Ease of starting, lower gas flow
		Argon-Helium	Helium addition improves penetration on heavy gage metal
	Manual welding	Argon	Better puddle control, especially for position welding
SS	Machine welding	Argon	Permits controlled penetration on thin gage material (up to 14 gage)
		Argon-Helium	Higher heat input, higher welding speeds possible on heavier gages
		Argon-Hydrogen (65%–35%)	Prevents undercutting, produces desirable weld contour at low current levels, requires lower gas flows
		Helium	Provides highest heat input and deepest penetration
Cu & Ni Cu–Ni Alloys (Monel & Inconel)		Argon	Ease of obtaining puddle control, penetration, and bead contour on thin gage metal
		Argon-Helium	Higher heat input to offset high heat conductivity of heavier gages
		Helium	Highest heat input for high welding speed on heavy metal sections
Ti		Argon	Low gas flow rate, minimum turbulence and air contamination of weld, improved metal transfer, improved heat-affected zone
		Helium	Better penetration for manual welding of thick sections (inert gas backing required to shield back of weld against contamination)
Si Bronze		Argon	Reduces cracking of this 'hot short' metal
Al Bronze		Argon	Less penetration of base metal

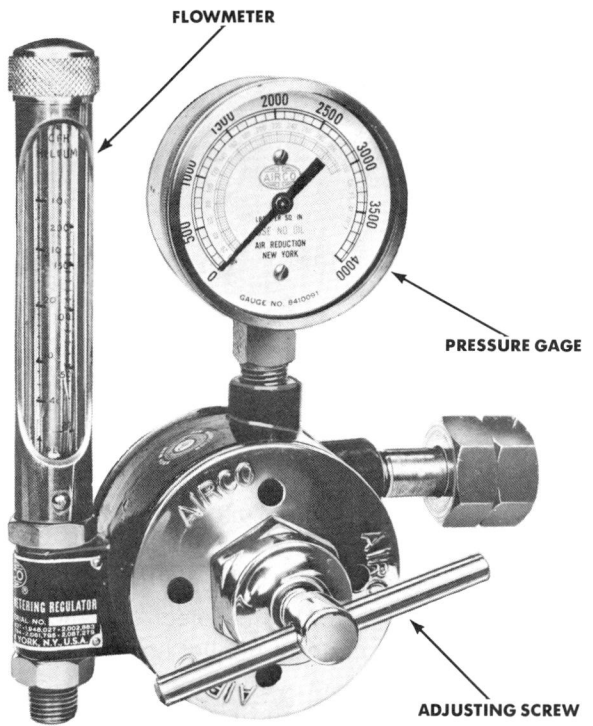

Fig. 11-6. An argon regulator with flowmeter. (Air Reduction Sales Co.)

specially designed regulator containing a flowmeter, shown in Fig. 11-6, may also be used. The advantage of the flowmeter is that it provides better gas flow control. The flowmeter is calibrated either to show the flow of gas in cubic feet per hour (cfh) or liters per minute (lpm).

The correct flow of argon to the torch is controlled by turning the adjusting screw on the regulator. The rate of flow required depends on the thickness of the metal to be welded.

Filler Rod

When Tig welding some metals a filler rod is required. Normally filler metal is not necessary on light gage materials since they can be made to readily flow together. Occasionally filler metal is added on thin pieces when it is essential to reinforce the joint.

Filler metal must be of the same composition as that of the base metal. Thus mild steel rods are used to weld low-carbon steel, aluminum rods for welding aluminum, copper rods for joining copper, and so on. Sometimes strips of the parent metal will serve as satisfactory filler metal.

Special filler rods are available for Tig welding. These rods are similar in classification to the filler wires used for Mig welding. See Unit 12. The copper-coated mild steel rods used for oxyacetylene welding are not suitable for Tig welding since they tend to contaminate the tungsten electrode. The special Tig filler rods contain a greater amount of deoxidizers, thereby producing less spattering in the weld and sounder weld joints.

In general, the diameter of the filler rod should be about the same as the thickness of the metal to be welded.

Protective Equipment

A helmet like the one used in metallic arc welding is required to protect the welder from arc radiation. The shade of the lens to be used depends upon the intensity of the arc.

Besides the regulation helmet, protective clothing such as an apron and gloves must be worn whenever welding with gas tungsten arc.

Backing the Weld

For many welding jobs, some suitable backing is necessary. On light gage metals, backing is used to protect the underside of the weld from atmospheric contamination and from burning through. On heavier stock, back-up bars draw some of the heat generated by the intense arc.

The type of material used for back-up bars depends on the metal to be welded. Copper bars are suitable for stainless steel. When welding aluminum or magnesium, steel or stainless steel back-up bars are needed.

The back-up bar should be designed so it does not actually touch the weld zone. See Fig. 11-7.

Welding Travel Direction

Either a backhand or forehand travel direction may be used in Tig welding. In backhand welding, the direction of travel is from left to right. In forehand welding travel is from right to left. See Fig. 12-11, page 122.

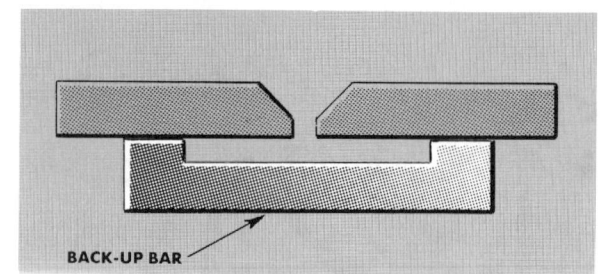

Fig. 11-7. A back-up bar should not touch the weld zone.

WELDING PROCEDURE

Preliminary Steps

Before starting to weld, follow these steps:

1. Check all electrical circuit connections to make sure they are tight.
2. Check for the proper diameter electrode and cup size. (Follow manufacturer's recommendations.)
3. Adjust the electrode so it extends about 1/8" to 3/16" beyond the end of the gas cup for butt welding and approximately 1/4" to 3/8" for fillet welding. See Fig. 11-8.
4. Check the electrode to make sure it is firmly held in the collet. Test it by placing the end against a solid surface and pushing the torch down gently but firmly. If the electrode moves into the nozzle, tighten the collet holder or gas cup. Be careful not to overtighten the gas cup because the threads will be stripped very easily.
5. Set the machine for the correct welding amperage.
6. If a water-cooled torch is to be used, turn on the water.
7. Turn on the inert gas and set the correct flow.

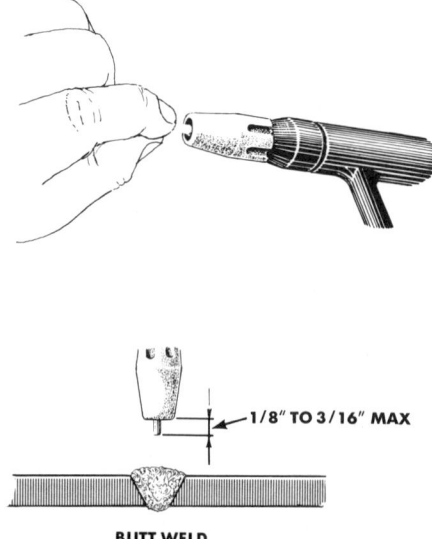

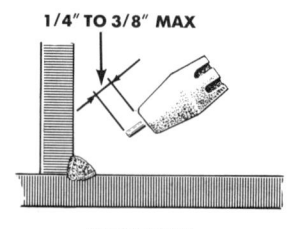

FILLET WELD

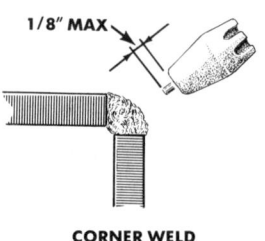

CORNER WELD

BUTT WELD

Fig. 11-8. Adjust the electrode so it extends beyond the edge of the gas cup.

Starting the arc. If you are using an AC machine, the electrode does not have to touch the metal to start the arc. To get the arc going, first turn on the welding current and hold the torch in a horizontal position about 2″ above the work. See Fig. 11-9. Swing the end of the torch toward the workpiece so the end of the electrode is 1/8″ above the plate, as in Fig. 11-10. The high-frequency current will jump the gap between the electrode and the plate, establishing the arc. *Be sure the downward motion is made rapidly* to provide the maximum amount of gas protection to the weld zone.

If a DC machine is used, hold the torch in the same position, but in this case the electrode must touch the plate to start the arc. When the

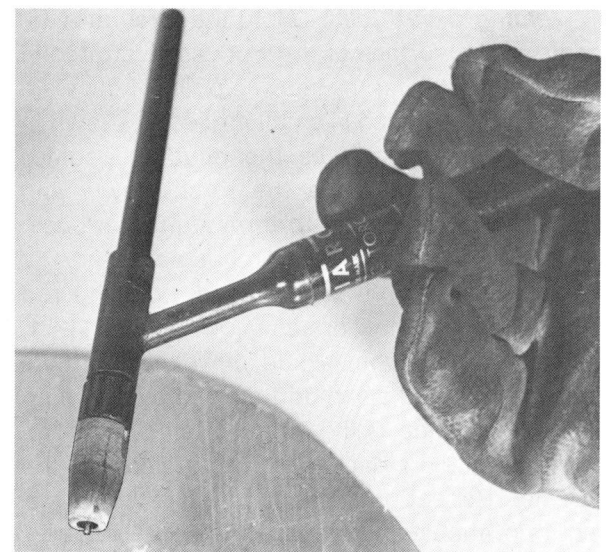

Fig. 11-9. To start the AC arc, first hold the torch in this manner.

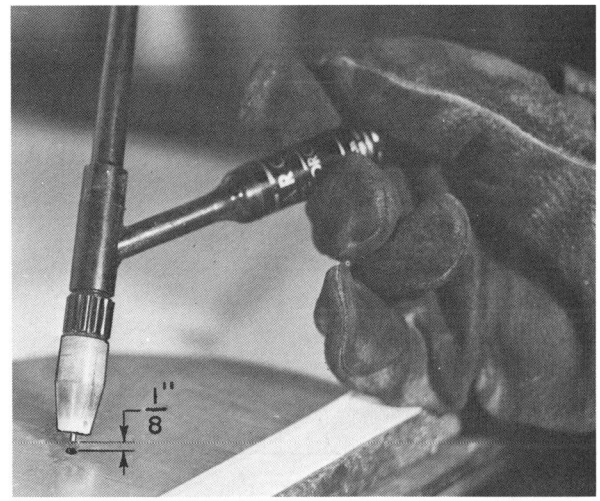

Fig. 11-10. Establish the AC arc by moving the tip of the electrode to within 1/8″ of the plate.

100 Arc Welding

arc is struck, withdraw the electrode so it is about 1/8" above the workpiece.

To stop the arc on the AC or DC machine, snap the electrode back to the horizontal position. Make this movement rapidly to avoid marring or damaging the weld surface.

If you are using a water-cooled cup, do not allow the cup to come in contact with the work when the current is on. The hot gases may cause the arc to jump from the electrode to the cup rather than the plate, thereby damaging the cup. Be sure that the water flow is set according to the manufacturer's recommendations.

Welding a butt joint. Hold the torch at a 75° angle to the surface of the work as illustrated in Fig. 11-11. Preheat the starting point of the weld by moving the torch in small circles as shown in Fig. 11-12. As soon as the puddle becomes bright and fluid, move the torch slowly and steadily along the joint to form a uniform bead. No circular motion of the torch is necessary.

If the filler rod is to be added, hold the rod about 15° from the work as shown in Fig. 11-11. As the puddle becomes fluid, move the arc to the rear of the puddle and add the rod by touching the leading edge of the puddle. Remove the rod and bring the arc back to the leading edge of the puddle. Repeat this sequence for the entire length of the seam. See Fig. 11-13.

Welding a lap and T-joint. To weld a lap or T-joint, first form a puddle on the bottom piece. After the puddle is formed, shorten the arc to about 1/16". Then rotate the torch directly over the

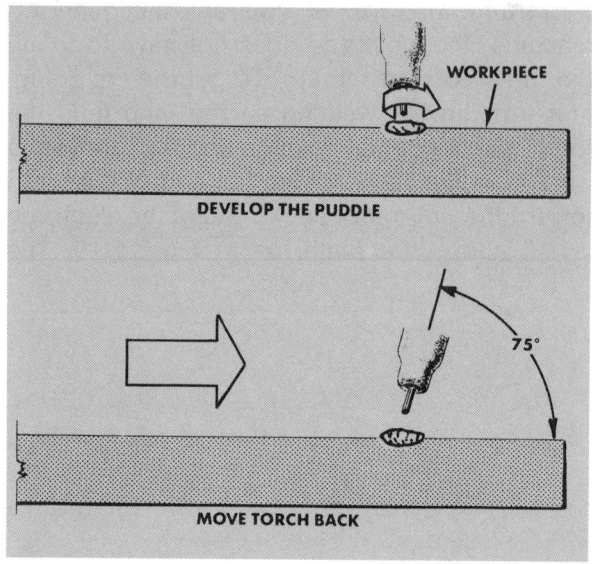

Fig. 11-12. Start the puddle by a circular movement of the torch. Once the puddle is formed, advance the puddle without circular motion.

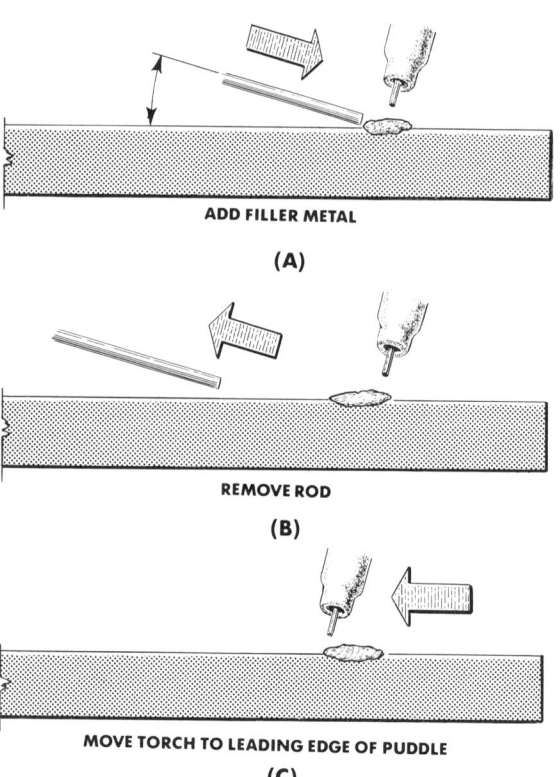

Fig. 11-13. Follow the steps in the top view in adding filler rod.

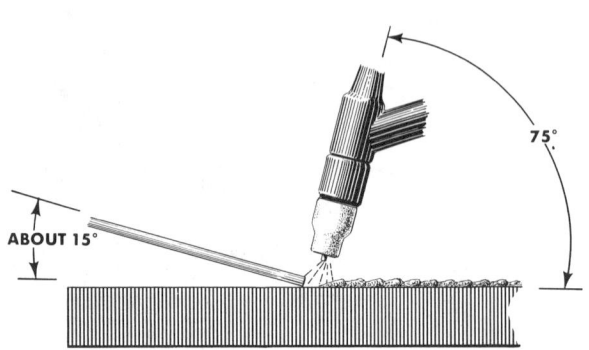

Fig. 11-11. Hold the torch at a 75° angle.

Gas Tungsten-Arc Welding

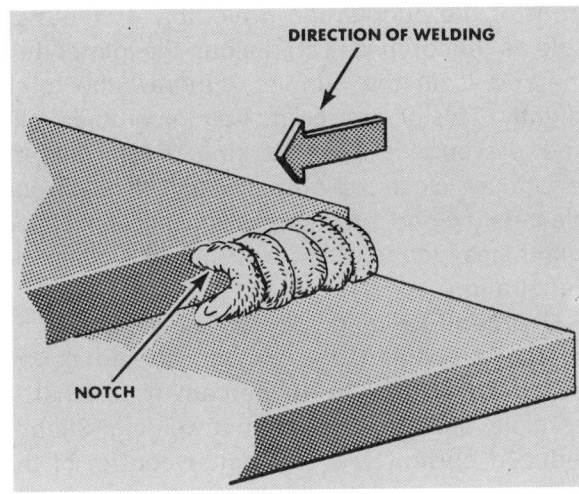

Fig. 11-14. Advance the torch just fast enough so the notch continues to form and move forward.

joint until the pieces are joined. After the welding is started, no further torch rotation is necessary. Move the torch along the seam with the end of the electrode just above the edge of the top sheet.

In welding a lap joint, you will find that the puddle forms a V shape. The center of the V is called a *notch,* and the speed at which this notch travels determines how fast the torch should be moved. Do not get ahead of it. See Fig. 11-14. Make certain that this notch is completely filled for the entire length of the seam. Otherwise there will be insufficient fusion and penetration.

If a filler rod is to be used on the lap joint, dip the end of the rod in and out of the puddle about every 1/4" travel of the puddle as shown in Fig. 11-15. Watch carefully to avoid laying bits of filler rod on the cold, unfused base metal. If you add just the right amount of rod at the correct moment, you will get a uniform bead of the proper proportions.

Welding a corner joint. A corner joint does not need any filler rod. Start the puddle at the beginning edge and move the torch straight along the seam. If you find that the molten metal has a tendency to roll off the edge, your speed is too slow. On the other hand, if the completed portion of the weld is rough and uneven, your speed is too fast.

Vertical welding. Vertical Tig welding on thin material is usually done in a downward position to achieve an adequate weld without burning through the metal. When a filler rod is to be used, add it from the bottom or leading edge of the puddle as shown in Fig. 11-16.

On heavier materials, an upward welding technique is preferred since deeper penetration can

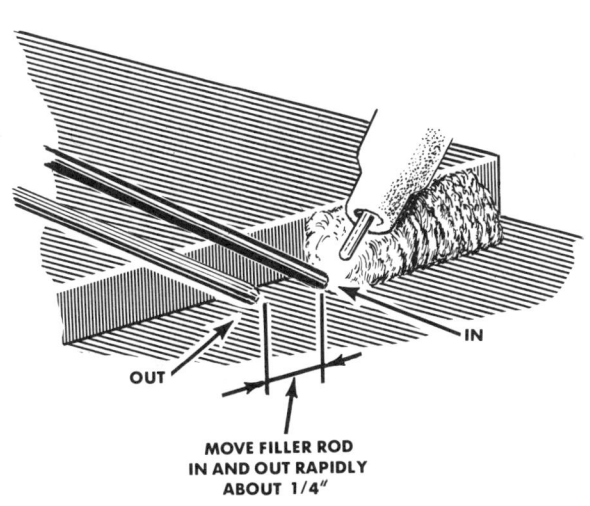

Fig. 11-15. Welding a lap joint with a filler rod.

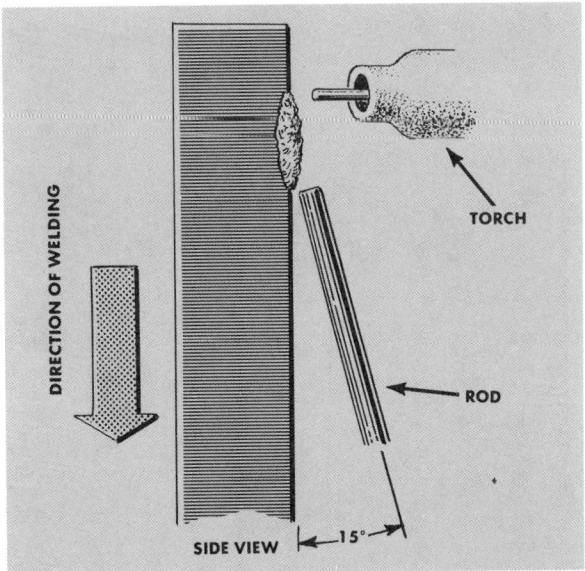

Fig. 11-16. Positioning the torch and rod for vertical downward welding.

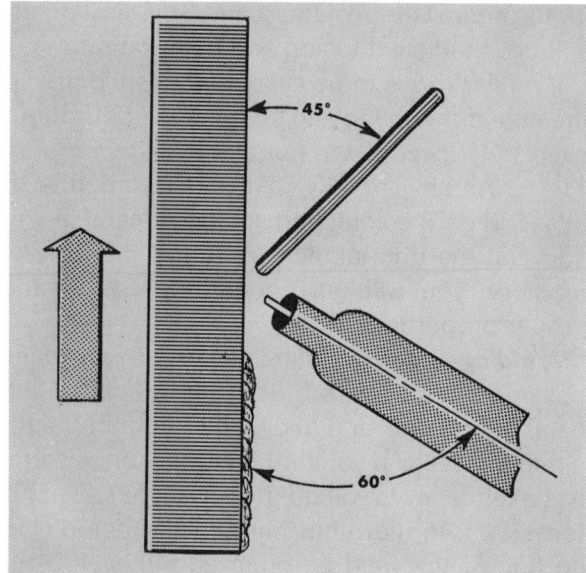

Fig. 11-17. Position of the torch and rod for vertical upward welding.

be achieved. Upward welding generally requires a filler rod. Notice in Fig. 11-17 the position of the torch and rod.

Horizontal welding. Start the arc about 1/2" from the joint. Once the arc is started, move it to the beginning of the joint. Hold the torch and rod as shown in Fig. 11-18. Dip the filler rod into the front of the puddle and preferably on the high side as the torch is moved along the joint. While the rod is in the puddle, withdraw the torch slightly. This allows the molten metal to solidify and prevents it from sagging. Keep the arc length as close as possible to the electrode diameter. Good arc length and correct speed eliminates undercutting and permits complete penetration.

Overhead welding. When Tig welding in an overhead position, the current should be reduced between 5 to 10 percent from what is normally used for flat position welding. Slightly reduced current will give better control of the weld puddle. Both the torch and the rod should be held as they would be in flat position welding. See Fig. 11-19. A smaller bead is advisable since it is less affected by the pull of gravity. Dip the filler rod in and out as in other welding positions. By pulling the arc back a little further, the filler metal will solidify faster.

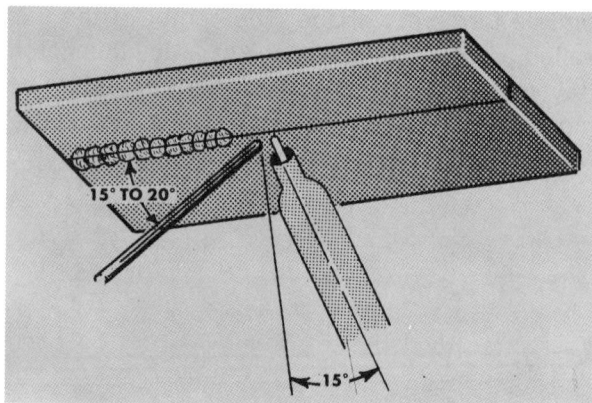

Fig. 11-19. Position of the torch and rod for Tig overhead welding.

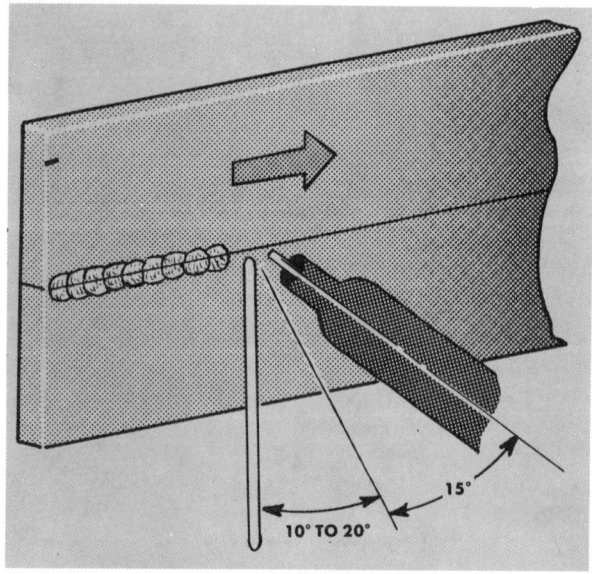

Fig. 11-18. Position of torch and rod for Tig horizontal welding.

TIG WELDING COMMON METALS

The actual technique of Tig welding various metals, such as aluminum, magnesium, copper, stainless steel, carbon steels and low-alloy steels, is virtually the same. In general, Tig produces better results on these metals than does the oxy-acetylene or shielded metal arc. Several specific welding characteristics for each of these metals are described in the paragraphs that follow.

torch should be moved in a straight line without a weaving motion.

Best results are obtained by using ACHF current with argon as a shielding gas. See Table 11-4 for recommended welding requirements of gas flow, current, etc.

TABLE 11-4. TIG WELDING—ALUMINUM.

STOCK THICKNESS (inches)	TYPE OF JOINT	AMPERES, AC CURRENT			ELECTRODE (inches) dia	ARGON FLOW 20 psi		FILLER ROD (inches) dia
		FLAT	HORIZONTAL & VERTICAL	OVERHEAD		lpm	cfh	
1/16	Butt	60–80	60–80	60–80	1/16	7	15	1/16
	Lap	70–90	55–75	60–80	1/16	7	15	1/16
	Corner	60–80	60–80	60–80	1/16	7	15	1/16
	Fillet	70–90	70–90	70–90	1/16	7	15	1/16
1/8	Butt	125–145	115–135	120–140	3/32	8	17	1/8
	Lap	140–160	125–145	130–160	3/32	8	17	1/8
	Corner	125–145	115–135	130–150	3/32	8	17	1/8
	Fillet	140–160	115–135	140–160	3/32	8	17	1/8
3/16	Butt	190–220	190–220	180–210	1/8	10	21	5/32
	Lap	210–240	190–220	180–210	1/8	10	21	5/32
	Corner	190–220	180–210	180–210	1/8	10	21	5/32
	Fillet	210–240	190–220	180–210	1/8	10	21	5/32

psi—pounds per square inch
lpm—liters per minute
cfh—cubic feet per hour

Aluminum. Non-heat-treatable wrought aluminum alloys in the 1000, 3000 and 5000 series are readily weldable. The heat-treatable alloys in the 2000, 6000, and 7000 series can be welded but higher welding temperatures and speeds are needed. Elimination of weld cracking in these alloys can often be achieved by using a rod that has a higher alloy content than the base metal.

While welding can be performed in any position, the task is simplified and the quality of the completed joint is more satisfactory if the weld is done in a flat position. Back-up blocks should be used wherever possible especially on plates 1/8" or less in thickness. In most cases, the

Stainless steel. Stainless steels, especially those in the 300 series, are very easy to Tig weld. Either direct-current straight-polarity or alternating current with high-frequency stabilization can be used. The gas tungsten-arc is particularly adaptable for welding light gage stainless steel. See Table 11-5 for specific welding requirements.

Carbon steels. Gas tungsten-arc is used extensively in welding low and medium-carbon and low-alloy steels because of the ease with which the welding can be accomplished and the greater protection from atmospheric contamination. For economic reasons gas tungsten-arc

TABLE 11-5. TIG WELDING—STAINLESS STEEL.

STOCK THICKNESS (inches)	TYPE OF JOINT	DC CURRENT STRAIGHT POLARITY amperes			ELECTRODE (inches) dia	ARGON FLOW 20 psi		FILLER ROD (inches) dia
		flat	horizontal & vertical	overhead		lpm	cfh	
1/16	Butt	80–100	70–90	70–90	1/16	5	11	1/16
	Lap	100–120	80–100	80–100	1/16	5	11	1/16
	Corner	80–100	70–90	70–90	1/16	5	11	1/16
	Fillet	90–110	80–100	80–100	1/16	5	11	1/16
3/32	Butt	100–120	90–110	90–110	1/16	5	11	1/16
	Lap	110–130	100–120	100–120	1/16	5	11	1/16
	Corner	100–120	90–110	90–110	1/16	5	11	1/16
	Fillet	110–130	100–120	100–120	1/16	5	11	1/16
1/8	Butt	120–140	110–130	105–125	1/16	5	11	3/32
	Lap	130–150	120–140	120–140	1/16	5	11	3/32
	Corner	120–140	110–130	115–135	1/16	5	11	3/32
	Fillet	130–150	115–135	120–140	1/16	5	11	3/32
3/16	Butt	200–250	150–200	150–200	3/32	6	13	1/8
	Lap	225–275	175–225	175–225	3/32	6	13	1/8
	Corner	200–250	150–200	150–200	3/32	6	13	1/8
	Fillet	225–275	175–225	175–225	3/32	6	13	1/8

TABLE 11-6. TIG WELDING—PLAIN CARBON AND LOW-ALLOY STEELS.

STOCK THICKNESS	DC CURRENT STRAIGHT POLARITY (amperes)	FILLER ROD dia (inches)	ARGON FLOW (psi)	
			lpm	chf
0.035	100	1/16	4–5	8–10
.049	100–125	1/16	4–5	8–10
.060	125–140	1/16	4–5	8–10
.089	140–170	1/16	4–5	8–10

welding is limited to materials under 1/4" in thickness. Ordinarily when filler rods are used they should contain sufficient deoxidizers to prevent porosity. Remember oxy-acetylene rods are not suitable for Tig welding. Table 11-6 includes specific welding requirements for Tig welding plain carbon and low-alloy steels.

SELF QUIZ

Correct answers are listed in the back of the book.

Fill-In

Supply the missing word or words where blanks appear.

1. In starting the arc you should first hold the torch over the workpiece in a _____ position and then swing the end of the torch so the electrode is about _____ above the plate.
2. In a DC machine the electrode must _____ the plate to start the arc.
3. To stop the arc the torch should be snapped back in a _____ position.
4. When welding in the flat position the Tig torch should be held at about a _____ degree angle.
5. In Tig welding the puddle is started by a _____ movement of the torch.
6. The filler rod should be added by touching the _____ of the puddle.
7. The gas mostly used in Tig welding is _____.
8. In Tig welding the electrode is used only to maintain the _____.

Short Answer

On the blanks provided write the answers to the following questions.

9. What does the abbreviation GTAW stand for? _____
10. What does ACHF mean? _____
11. What does DCRP mean? _____
12. Suppose that DCRP is used for Tig Welding. Will the electrode diameter be larger or smaller than that used with DCSP? _____
13. Does DCSP or DCRP produce a deeper and narrower weld? _____
14. In Tig welding, is straight or reverse current best for welding aluminum? _____
15. Is AC or DC current usually better for welding aluminum and magnesium? _____
16. What do the following abbreviations mean?
    A. cfh   A. _____
    B. lpm   B. _____

True and False

Circle the letter T if the statement is true or the letter F if it is false.

17. T F Straight polarity with DC current is best for Tig welding aluminum.
18. T F The use of copper back-up blocks is often advisable when welding thin gage aluminum.
19. T F Once an arc is started with a DC machine, the electrode should continue to rest lightly on the metal plate.
20. T F The rate of gas flow should be about the same for welding most metals.
21. T F When using a water-cooled torch, the cup should not be allowed to come in contact with the plate when the current is on.
22. T F The tungsten electrode for ACHF Tig aluminum welding should have a sharp point.
23. T F ACHF has better cleaning action for removing oxide on aluminum than DCSP.
24. T F Once a puddle is formed in Tig welding, the puddle is usually advanced without any circular torch motion.
25. T F When using a filler rod in Tig welding,

the rod should be held at the same angle as the torch.
26. T F The filler rod should be kept constantly in the molten puddle.
27. T F DC straight polarity current should be used in Tig welding low carbon steel.
28. T F A filler rod that is normally used for gas welding is suitable for Tig welding as well.
29. T F With an AC machine the electrode has to touch the metal plate to start the arc.
30. T F The downward motion of the torch with an AC machine is executed rapidly to prevent the electrode from sticking to the plate.
31. T F Welding with DCRP produces deep penetration because it concentrates heat at the plate.
32. T F Welding with DCRP produces good cleaning action, but the weld penetration is shallow.
33. T F Welding with ACHF produces good cleaning action and deep weld penetration.
34. T F Air-cooled torches are used for welding light gage materials where low current values are required.
35. T F Water-cooled torches are used primarily on AC equipment.
36. T F The function of the gas cup on a Tig torch is to provide distributional control for the shielding gas.
37. T F A control switch on the torch starts the flow of both current and gas.
38. T F Pure tungsten electrodes are considered superior to the alloyed type, because they last longer.

WELDING ASSIGNMENT

I. Depositing Straight Beads

1. Get a piece of low carbon steel plate about 1/8" in thickness and draw a series of straight lines on it. See Fig. 11-20.

2. Use a 1/16" electrode and a suitable filler rod. Set the machine at about midpoint between 125-140 amperes and at a gas (argon) flow of 8-10 cfh.

3. First just practice starting and stopping the arc. Once you can do this readily, melt a small puddle and carry it forward without using any filler rod. Next run straight layers of beads over the lines using a filler rod. Experiment with the amperage setting by raising and lowering it from the original setting and notice the effects this has in depositing uniform beads. Be sure to hold the filler rod near the leading edge of the puddle and dip it in and out of the puddle as you carry the beads forward. Do not use any torch weaving motion.

4. Continue to deposit straight beads until your instructor certifies that you have mastered this skill.

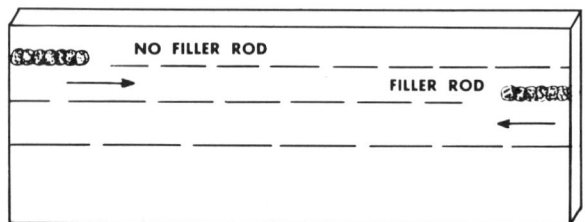

Fig. 11-20. Practice plate for depositing surface beads with Tig.

Gas Tungsten-Arc Welding **107**

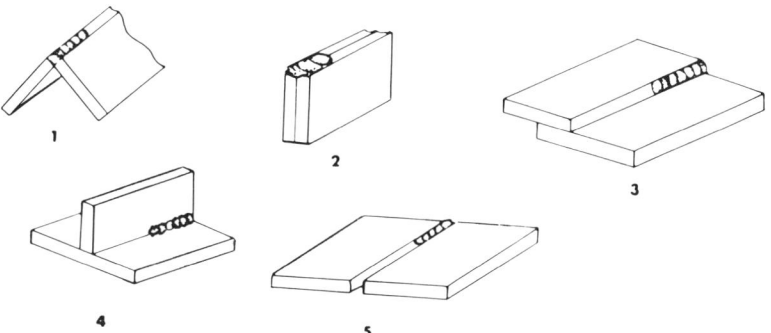

Fig. 11-21. Sequence of practice Tig welds.

II. Welding Joints with Tig in a Flat Position

1. Secure several mild steel plates about 0.06" in thickness. Tack weld them to form corner, edge, lap, T and butt joints.

2. First practice welding corner and edge joints without a filler rod. See Fig. 11-21.

3. Next practice welding lap joints with a filler rod and depositing a single pass. See Fig. 11-21.

4. Practice welding T joints with a filler rod and using one pass. See Fig. 11-21.

5. On butt joints, start with a square edge making a single pass with a filler rod. See Fig. 11-21.

6. Test each joint as it is completed by bending it in a vise. Use the Weld Analysis Check List, shown at the end of this unit, to evaluate the soundness of your welds.

7. Continue to practice welding joints in the flat position until your instructor tells you that you are ready to go on to the next assignment.

III. Welding Joints with Tig in a Vertical Position

1. Secure several pieces of light gage low carbon steel plates about $1/16$" in thickness. Bend a small flange on the edges at a height equal to approximately the thickness of the plates. See Fig. 11-22. Tack weld them to form butt joints.

2. Place the workpiece in a suitable jig for vertical welding. Make a downward weld without using a filler rod.

3. Continue downward welding more such joints until you have mastered this technique.

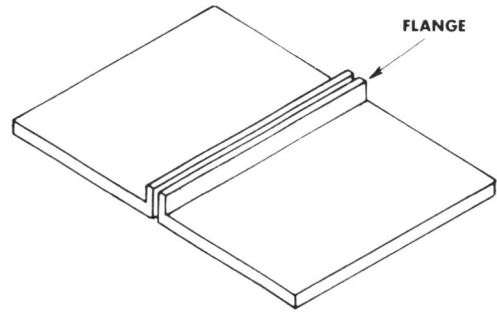

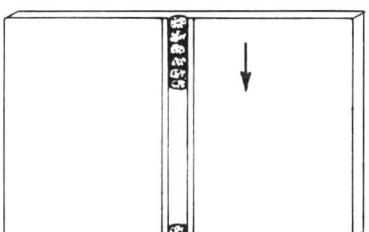

Fig. 11-22. Practice plates for Tig downward welding.

4. Test each welded joint by bending it and evaluate the results using the Weld Analysis Check List shown at the end of this unit.

IV. Welding Joints with Tig in a Horizontal Position

1. Secure several pieces of mild steel plates about 0.089" in thickness and tack weld them to form lap and butt joints.

2. Fasten the workpiece in a suitable jig for

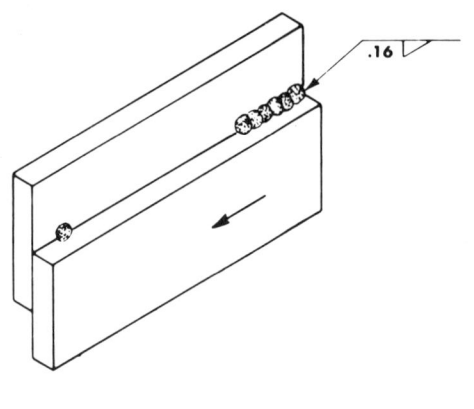

Fig. 11-23. Practice plates for horizontal Tig welding.

horizontal welding and practice welding each joint with a single pass. See Fig. 11-23. Be sure your amperage and gas flow are properly set for welding this metal thickness.

3. Test each weld by bending it in a vise. Use the Weld Analysis Check List at the end of this unit to evaluate your welds.

V. Welding Joints with Tig in an Overhead Position

1. Unless informed otherwise by your teacher, practice overhead Tig welding on two types of joints—butt and lap. Use either 0.60" or 0.125" low carbon steel plates.

2. Tack weld the pieces to form butt and lap joints and clamp in a jig for overhead welding. See Fig. 11-24.

3. Continue to practice overhead welding until you have mastered this skill. Test each weld and evaluate the results by using the Weld Analysis Check List shown at the end of this unit.

VI. Tig Welding Aluminum

1. Weld corner, lap, T and butt joints. Use available materials ranging in thickness from 0.063" to 0.125". Practice welding these joints in the positions shown in Fig. 11-25.

2. Use an aluminum filler rod of the right diameter. Be sure to dip the rod in and out of the puddle as the weld moves forward. Do not melt the rod with the arc, because if you do it will ball up in front of the puddle.

3. Follow the same welding procedure as you did in welding mild steel. Strike the arc a short distance ahead where the weld is to begin and

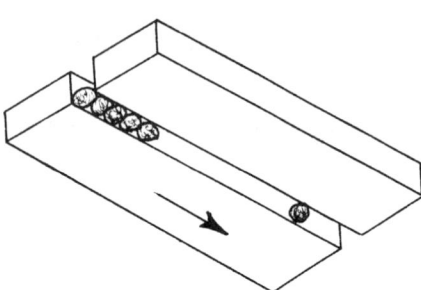

Fig. 11-24. Practice plates for overhead Tig welding.

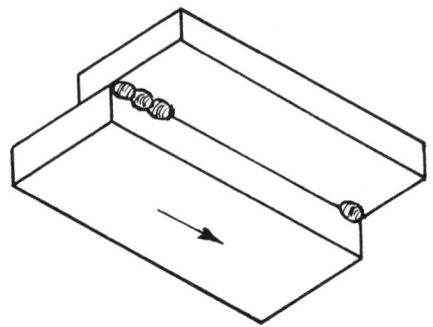

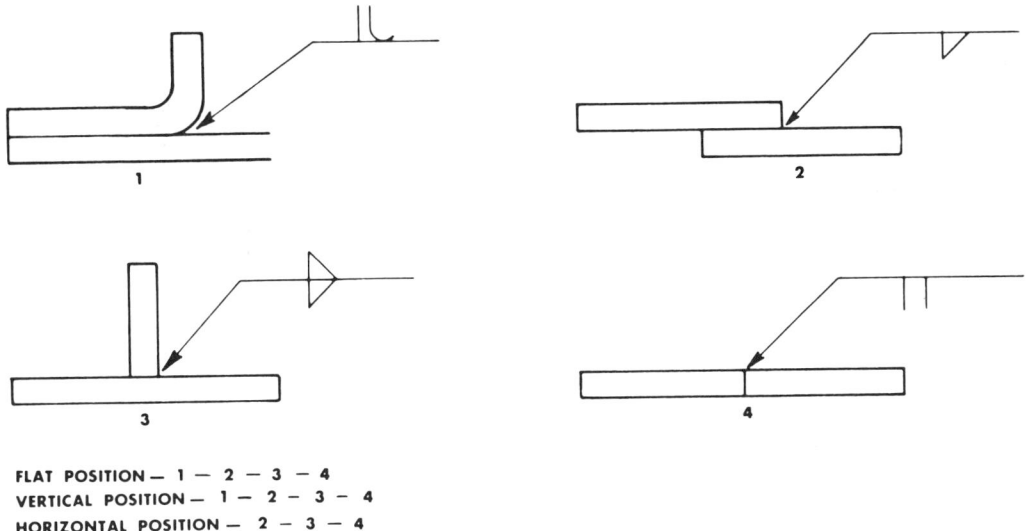

Fig. 11-25. Aluminum Tig welding sequence.

then bring it back to the starting point. Keep the torch steady and move it forward from right to left in a straight line without any weaving motion. This will help you deposit uniform beads with evenly spaced close ripples. By inclining the torch as previously described in your book, you will be able to see the welding process better, and you will have greater puddle control. Remember, when aluminum melts, the puddle does not behave like the running puddle of mild steel. Before you even begin to weld any of the joints, just play the arc on the surface of a scrap piece and see what happens when the melting point is reached.

4. Test each weld as it is completed and evaluate the results as you have done in previous assignments.

VII. Tig Welding Stainless Steel

1. Practice flat position welding on corner, edge, lap, T and butt joints. See Fig. 11-26. Use stock thickness prescribed by your instructor. Make all welds with a single pass. Use filler rod on all joints except on the flange butt joint. Be sure the filler rod is of a stainless steel quality suitable for the metal you are welding. Select the current and gas flow recommended for the stock

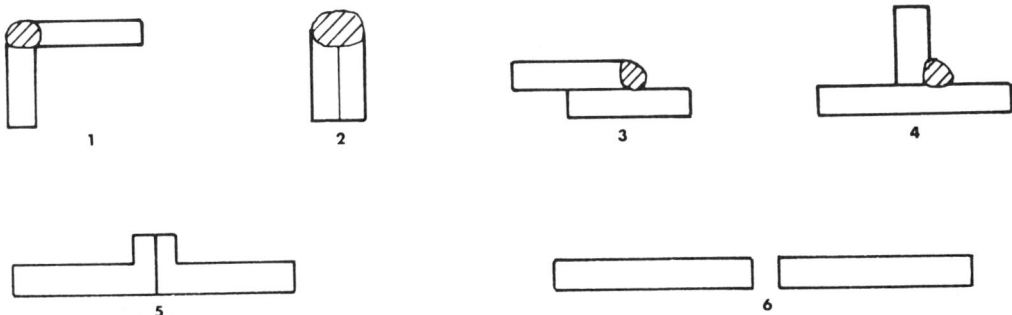

Fig. 11-26. Sequence of stainless steel practice welds in flat position.

110 Arc Welding

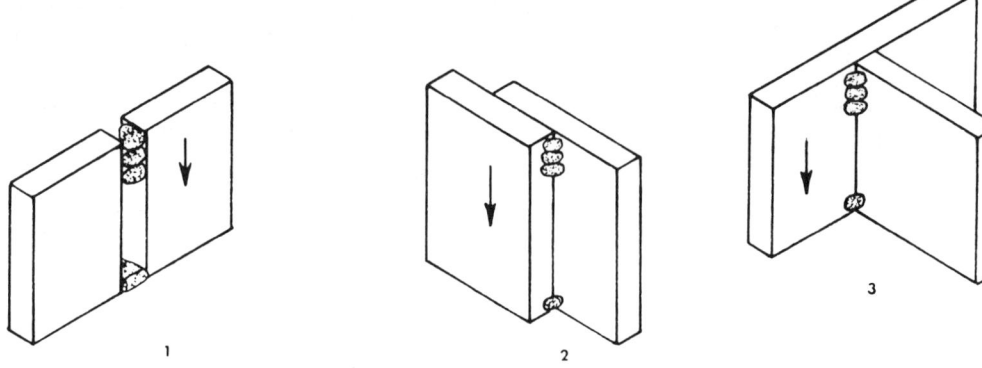

Fig. 11-27. Sequence of stainless steel practice welds in vertical position.

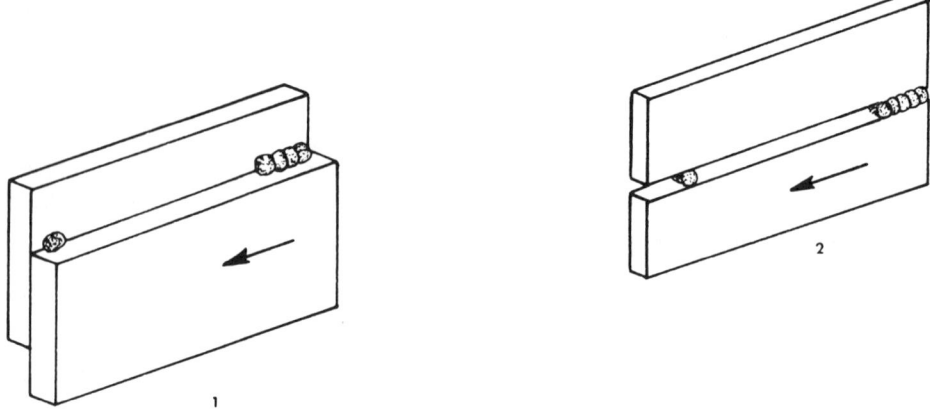

Fig. 11-28. Sequence of stainless steel practice welds in horizontal position.

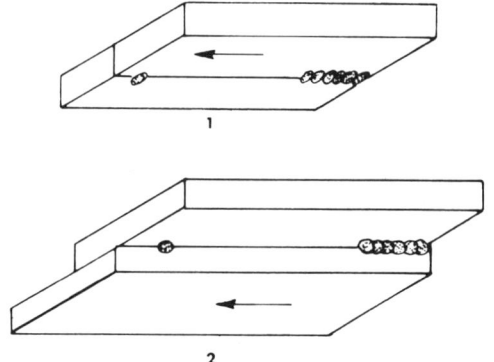

Fig. 11-29. Sequence of stainless steel practice welds in overhead position.

thickness and weld position. Use the forehand welding technique—right to left.

2. Practice welding butt, lap and T joints in vertical position. See Fig. 11-27.

3. Practice welding lap and butt joints in horizontal position. See Fig. 11-28.

4. Practice butt and lap welds in overhead position. See Fig. 11-29.

5. Test each weld as you complete it and always evaluate your results with the help of the Weld Analysis Check List shown at the end of this unit.

Weld Analysis Check List

		Yes	No
1.	Bead widths are right size	___	___
2.	Beads have uniform ripples	___	___
3.	Weld beads are too flat	___	___
4.	Weld beads are too high	___	___
5.	Weld penetration is insufficient	___	___
6.	Weld penetration is excessive	___	___
7.	Cold laps on surface	___	___
8.	Weld has surface porosity	___	___
9.	Weld has subsurface porosity	___	___
10.	Weld has crater cracks	___	___
11.	Weld has burn thru	___	___
12.	End crater is filled	___	___
13.	Weld passed bend test without cracking	___	___

UNIT 12
Gas Metal-Arc Welding

The Gas Metal Arc welding process (Mig), sometimes referred to as GMAW, uses a continuous, consumable wire electrode. The molten weld puddle is completely covered with a shield of gas. The wire electrode is fed through the torch at pre-set controlled speeds. The shielding gas is also fed through the torch. See Fig. 12-1.

Welding Current

Different welding currents have a large effect on the results obtained in gas metal-arc welding. Optimum efficiency is achieved with direct current reverse polarity (DCRP). The heat in this instance is concentrated at the weld puddle and therefore provides deeper penetration at the weld. Furthermore, with DCRP there is greater surface cleaning action which is important in welding metals having heavy surface oxides, such as aluminum and magnesium.

Straight polarity (DCSP) is impractical with Mig welding because weld penetration is wide and shallow, spatter is excessive, and there is no surface cleaning action. The ineffectiveness of straight polarity largely results from the pattern of metal transfer from the electrode to the weld puddle. Whereas in reverse polarity the transfer is in the form of a fine spray, in straight polarity the transfer is largely of the erratic globular type. The use of AC current is never recommended since the burn-offs are unequal on each half cycle.

Types of Metal Transfer

When welding with consumable wire electrodes, the transfer of metal is achieved either by the spray transfer or by the short-circuiting transfer methods. The type of metal transfer that occurs will depend on electrode wire size, shielding gas, arc voltage, and welding current.

Spray transfer. In spray transfer very fine droplets or particles of the electrode wire are rapidly projected through the arc from the end of the electrode to the workpiece in the direction in which the electrode is pointed. The droplets are equal to or smaller than the diameter of the electrode. While in the process of transferring through the welding arc, the metal particles do not interrupt the flow of current and there is a

Gas Metal-Arc Welding 113

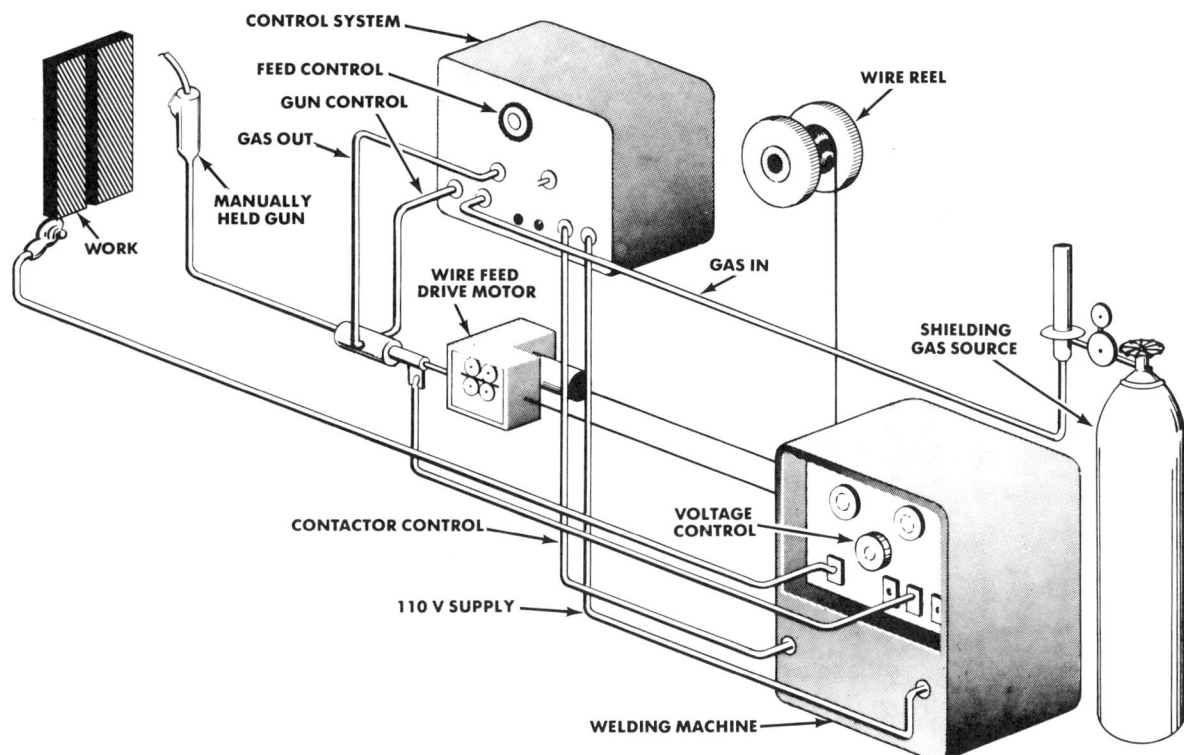

Fig. 12-1. Mig welding unit. (Hobart Brothers Co.)

nearly constant spray of metal. The use of argon or a mixture of argon and oxygen is also necessary for spray transfer. Argon produces a pinching effect on the molten tip of the electrode, permitting only small droplets to form and transfer during the welding process. See Fig. 12-2. Since the individual drops are small, the arc is stable and can be directed where required. Because the metal transfer is produced by directional force, which is stronger than gravity, spray transfer is effective for out-of-position welding. It is particularly suitable for welding heavy gage metal. It is not too practical for welding light gage metal because of the resulting burn-through.

Short circuiting transfer (short arc). The short circuiting transfer permits welding thinner sections with greater ease. It is extremely practical for welding in all positions, especially in

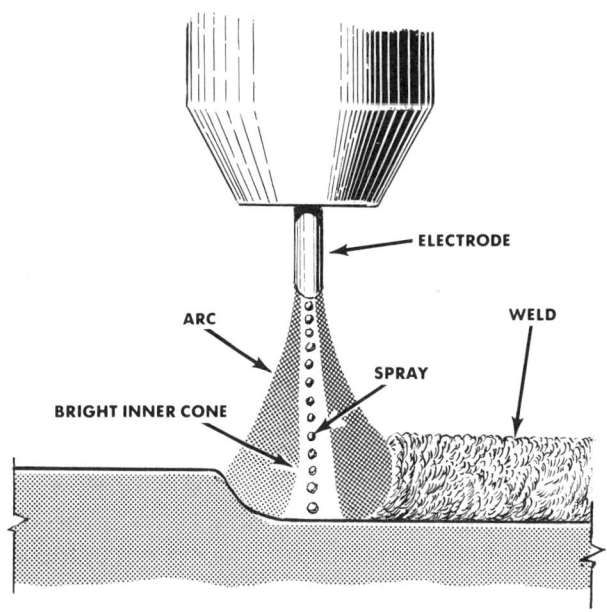

Fig. 12-2. Spray metal transfer.

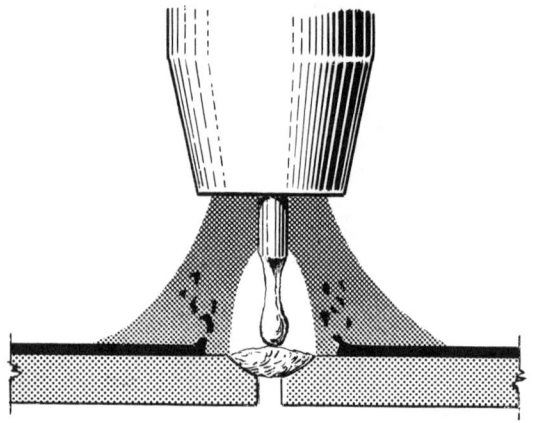

Fig. 12-3. Electrode makes contact with workpiece, creating a short circuit. Arc is extinguished and allowed to cool. Frequency of arc extinction in SHORT ARC varies from 20 to 200 times per second, depending on job requirements. (Linde Co.)

vertical, horizontal and overhead welding where normally puddle control is a little difficult.

With this process a shallow weld penetration is obtained. See Fig. 12-3. It is generally considered to be the most practical at current levels below 200 amperes with fine wire of 0.045" or less in diameter. The use of fine wire produces weld pools that remain relatively small and are easily managed, making all-position welding possible.

As the molten wire is transferred to the weld, each drop touches the weld puddle before it has broken away from the advancing electrode wire. The circuit is shorted, and the arc is then extinguished.

Electromagnetic pinch force squeezes the drop from the wire. The short circuit is broken and the arc re-ignites. Shorting occurs from 20 to 200 times a second depending on preset controls. Shorting of the arc pin-points the effective heat. The result is a small, relatively cool weld puddle which reduces burn-through.

In *short-arc welding,* the shielding gas mixture consists of 25 percent carbon dioxide, which provides increased heat for higher speeds, and 75 percent argon which controls spatter. However, considerable usage is now being made of straight CO_2 where bead contour is not particularly important but good penetration is very essential.

Power Supply

The recommended machine for Mig welding is a rectifier or motor generator supplying direct current with normal limits of 200 to 250 amperes for all position welding. Direct current reverse polarity (DCRP) is used for optimum efficiency. DCRP contributes to better melting, deeper penetration, and excellent cleaning action.

In Mig welding, heat is generated by the flow of current through the gap between the end of the wire electrode and the workpiece. A voltage forms across this gap which varies with the length of the arc. To produce a uniform weld, the welding voltage and arc length must be maintained at a constant value. Since the older welding machines could not adequately control those two things, a new machine, known as the *constant voltage (potential) power welder,* was developed. With this machine, the voltage is preset and remains constant, regardless of the amount of current drawn. Thus the power supply becomes self-correcting with respect to arc length. The arc length can be set on the power supply and any variations in nozzle-to-work distance will not produce changes in the arc length. For example, if the arc length becomes shorter than the pre-selected value, there is an automatic increase of current and the wire speed automatically adjusts itself to maintain a constant arc length. Similarly, if the arc becomes too long, the current decreases and the wire begins to feed faster.

Stated in another way, when the wire is fed into the arc at a specific rate, a proportionate amount of current is automatically drawn. The constant potential welder therefore provides the necessary current required by the load imposed on it. When the electrode wire is fed faster, the current increases; if it is fed slower, the current decreases.

Because of this self-correcting feature, less skill is necessary to get good welds. There are only two basic controls: a rheostat on the welding machine to regulate the voltage and a rheostat on the wire feed mechanism to control the speed of the wire feed motor.

There is no amperage control on a constant voltage type machine; the welding current output is determined by the wire feeder.

Wire Feeding Mechanism

The wire feeding mechanism automatically drives the electrode wire from the wire spool to the gun and arc. See Fig. 12-4. Control on the panel can be adjusted to vary the wire feeding speed. In addition, the control panel usually includes a welding power contactor and a solenoid to energize the gas flow. On units designed for welding with a water-cooled gun, a control is also available to turn on and shut off the water flow.

The wire feeder can be mounted on the power supply machine or it can be separate from the welding machine and mounted elsewhere to facilitate welding over a large area.

Welding Gun

The function of the welding gun is to deliver the wire, shielding gas, and welding current to the arc area. The manually operated gun is either water or air-cooled. An air-cooled gun is especially designed to weld light gage metals that require less than 200 amperes with argon as a shielding gas. However, such a torch can usually function at higher amperage (300A) with CO_2 because of the cooling effects of this gas. A water-cooled gun is generally best when welding with currents that are higher than 200 amperes.

Guns are either of the push or pull type. The pull gun has drive rolls that pull the welding wire from the wire feeder, and the push gun has the wire pushed to it by drive rolls in the wire feeder itself. The pull gun handles small diameter wires, while the push gun moves heavier diameter wires. The pull type is also used to weld with soft wires, such as aluminum and magnesium, while the push gun is considered more suitable for welding with hard wires, such as carbon and stainless steels, and currents in excess of 250 amperes.

Both guns have a trigger switch that controls the wire feed and arc as well as the shielding gas and water flow. When the trigger is released the

Fig. 12-4. Typical wire feeding unit for Mig welding. (Miller Electric Manufacturing Co.)

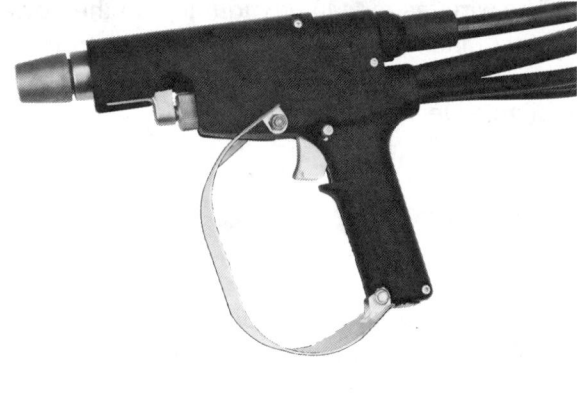

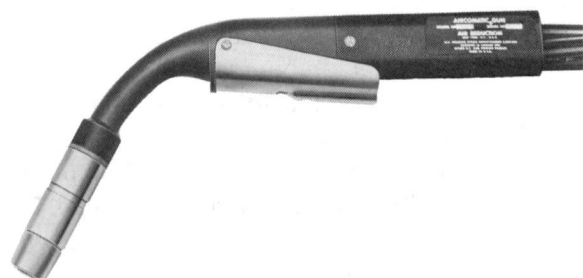

Fig. 12-5. Types of guns used for Mig welding.

wire feed, arc, shielding gas, and water, if a water-cooled torch is used, stop immediately. With some equipment a timer is included to permit the shielding gas to flow for a predetermined time to protect the weld until it solidifies.

Guns are available with a straight or curved nozzle. See Fig. 12-5. The curved nozzle provides easy access to intricate joints and difficult-to-weld patterns.

Shielding Gas

In any gas shielded arc welding process, the shielding gas can have a large effect upon the properties of a weld deposit. Therefore, welding is done in a controlled atmosphere. In shielded metal-arc welding, this is accomplished by placing a coating on the electrode which produces a non-harmful atmosphere when it disintegrates in the welding arc. In the case of Mig welding, the same effect is accomplished by surrounding the arc area with gases supplied from an external source.

The air in the arc area is displaced by the shielding gas. The arc is then struck under the blanket of shielding gas and the welding is accomplished.

The gas used for Mig welding depends on the kind of metal to be welded. For some operations straight argon is recommended. For other purposes a mixture of argon and oxygen is preferred. In some cases carbon dioxide or a mixture of oxygen and helium produces the best results. Table 12-1 lists the various gases recommended for welding different metals.

Gas Flow and Regulation

For most welding conditions, the gas flow rate will approximate 35 cu ft per hour. This flow rate may be increased or decreased depending upon the particular welding application. (See Tables 12-4, 12-5, 12-6, 12-7.)

The data presented in these tables are not intended as absolute settings but only as a point in making the starting settings. Final adjustments must often be made on a trial and error basis. Actually the correct settings will be governed by the type and thickness of metal to be welded, position of the weld, kind of shielding gas used, diameter of electrode and type of joint.

The proper amount of gas shielding usually results in a rapidly crackling or sizzling arc sound. Inadequate gas shielding will produce a popping arc sound with resultant weld discoloration, porosity, and spatter.

Gas drift may occur with high weld travel speeds or with unusually drafty or windy conditions in the weld area. Since one or more of these factors may cause the gas to drift away from the arc, the result is inadequate gas coverage. See Fig. 12-6. The gas nozzle should be adjusted for proper coverage and outside influences should be eliminated by proper windbreakers or shields. See Fig. 12-7.

TABLE 12-1. SHIELDING GASES FOR MIG WELDING.

MATERIAL	PREFERRED GAS	REMARKS
Aluminum alloys	Argon	With DC reverse polarity removes oxide surface on work piece
Magnesium aluminum alloys	75% He 25% A	Greater heat input reduces porosity tendencies. Also cleans oxide surface
Stainless steels	Argon + 1% O_2	Oxygen eliminates under-cutting when DC reverse polarity is used
	(Argon + 5% O_2)	When DC straight polarity is used 5% O_2 improves arc stability
Magnesium	Argon	With DC straight polarity removes oxide surface on work piece
Copper (deoxidized)	75% He, 25% A	Good wetting and increased heat input to counteract high thermal conductivity. Light gages
	(Argon)	
Low-carbon steel	Argon + 2% O_2	Oxygen eliminates under-cutting tendencies also removes oxidation
Low-carbon steel	Carbon dioxide (spray transfer)	High quality low current out of position welding low spatter
	Carbon dioxide (buried arc)	High speed low cost welding accompanied by spatter loss
Nickel	Argon	Good wetting decreases fluidity of weld metal
Monel	Argon	Good wetting decreases fluidity of weld metal
Inconel	Argon	Good wetting decreases fluidity of weld metal
Titanium	Argon	Reduces heat-affected zone, improves metal transfer
Silicon bronze	Argon	Reduces crack sensitivity of this hot short material
Aluminum bronze	Argon	Less penetration of base metal Commonly used as a surfacing material

Note: ( ) = Second choice

Correct positioning of the nozzle with respect to the work will be determined by the nature of the weld. The gas nozzle may usually be placed up to 2" from the work. Too much space between nozzle and work reduces the effectiveness of a gas shield while too little space may result in excessive weld spatter which collects on the nozzle and shortens its life.

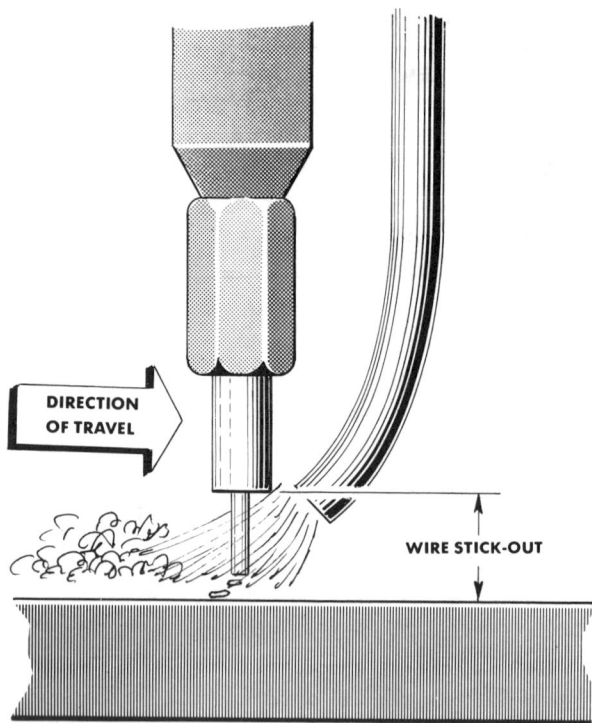

Fig. 12-6. Gas drift with inadequate gas coverage. (Hobart Brothers Co.)

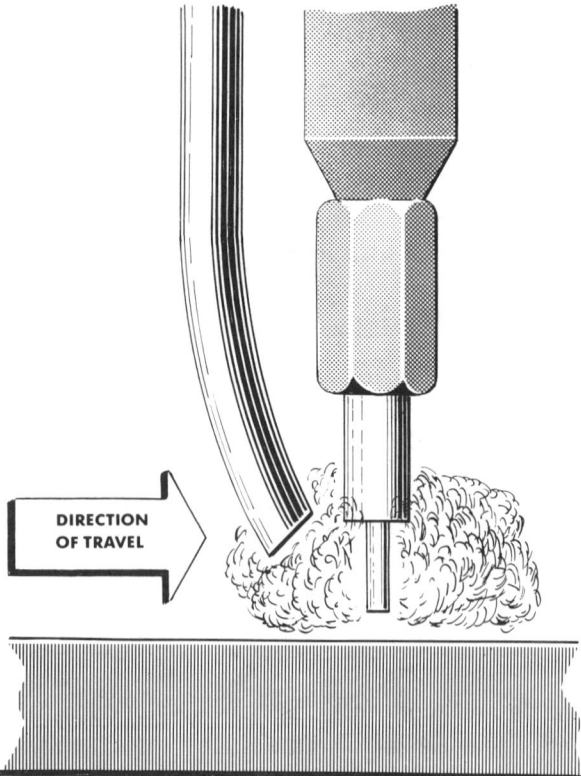

Fig. 12-7. Adequate gas coverage. (Hobart Brothers Co.)

Wire for Mig Welding

Filler wire for Mig welding should be similar in composition to the base metal. Several common wires and their suitability for a particular type of welding are listed in Table 12-2. These designations are based on the AWS classification system. Thus for mild steel wires, the *E* identifies it as an electrode, the next two digits show the tensile strength in psi per thousand, the *S* indicates a solid bare wire, and the final symbols specify a particular classification based on the chemical composition of the wire.

Wires are usually available in spools of several different sizes as well as in 36" rod lengths for Tig welding.

Best results are obtained by using the proper diameter wire for the thickness of the metal to be welded and the position in which the welding is to be done. See Tables 12-4, 12-5, 12-6, 12-7.

Basic wire diameters are 0.020", 0.030", 0.035", 0.045", 1/16" and 1/8". Generally wires of 0.020", 0.030" or 0.035" are best for welding thin metals. These wires are sometimes referred to as *micro wires*. The use of micro wires permits increased welding speeds and improves the appearance and quality of the welds. Micro-wire welding is especially adaptable for joining thin materials (20 gage), although it also can be used to weld low and medium-carbon steels, and low-alloy/high-strength steels of medium thickness. Medium thickness metals normally require 0.045" or 1/16" diameter electrodes. For thick metals, 1/8" electrodes are usually recommended. However, the position of welding is a factor which must be considered in electrode selection. Thus for vertical or overhead welding, smaller diameter electrodes will be more satisfactory than larger diameter wires.

Wire feed. The amperage of the welding current used limits the speed of the wire feed to a definite range. However, it is possible to make adjustments of the wire feed within the range. For a specific amperage setting, a high speed of wire feed will result in a short arc. A low speed contributes to a long arc. Also, the speed for overhead welding must be higher than that for flat position welding. (See Tables 12-4, 12-5, 12-6, 12-7.)

TABLE 12-2. FILLER WIRES FOR GAS-SHIELDED ARC WELDING

	MILD STEEL WIRES
E-60S-1	Silicon deoxidized wire for low and medium-carbon steels. Can be used either with CO_2, argon, or argon-CO_2 mixtures. Performs best on killed steels
E-60S-2	Premium quality wire containing Al, Zr, and Ti in addition to silicon and manganese deoxidizers. Can be used with CO_2 or argon-CO_2 or argon-O_2. Recommended for pipe welding and heavy vessel construction
E-60S-3	Used for higher quality welding either with CO_2, argon-O_2, or argon-CO_2 mixtures. Produces medium quality welds in rimmed steels and high quality welds in semi-killed steels
E-70S-1B	Low-alloy wire for carbon steels, low-alloy steels, and high strength low-alloy steels
E-70S-3	General purpose welding of low to medium-carbon steels. Has a silicon content high enough to permit its use in either CO_2, argon-oxygen mixtures or mixture of the two
E-70S-6	Contains higher manganese and silicon levels and has more powerful deoxidizing characteristics for welding over rust and scale or where stringent cleaning practices cannot be followed
E-70S-5	Contains aluminum and is designed for single or multipass welding of rimmed, semikilled, or killed mild steels. Suitable to weld steels having rusty or dirty surfaces and normally used with CO_2 gas
	ALUMINUM WIRES
ER-1100 ER-4043 ER-5183 ER-5554, 5556 ER-5654	To weld aluminum of similar composition
	STAINLESS STEEL WIRES
ER-308L	For welding types 304, 308, 321, 347
ER-308L-Si	For welding types 301, 304
ER-309	For welding types 309 and straight chromium grades when heat treatment is not possible. Also for 304-clad
ER-310	For welding types 310, 304-clad and hardenable steels
ER-316	For welding 316
ER347	For welding types 321, and 347 where maximum corrosion resistance is required
	COPPER AND COPPER-BASE ALLOY WIRES
E-CuSi (Silicon Bronze) E-CuAl-A1 (Aluminum Bronze) E-Cu (Deoxidized Copper) E-CuAl-A2 (Aluminum Bronze) E-CuAl-B (Aluminum Bronze)	Special wires for welding copper and copper-based alloys

Wire stick-out. Wire stick-out refers to the distance the wire projects from the nozzle of the gun. See Fig. 12-8, *top.* Stick-out influences the welding current since it changes the preheating in the wire. When the stick-out increases the preheating increases, which means that the power source does not have to furnish as much welding current to melt the wire at a given feed rate. Since the power source is self-regulating, the current output is automatically decreased. Conversely, if the stick-out decreases, the power source is forced to furnish more current to burn off the wire at the required rate.

For most Mig welding applications, the wire stick-out should measure from 3/8″ to 3/4″. On micro wires a shorter wire stick-out ranging from 1/4″ to 3/8″ is recommended. An excessive amount of wire stick-out results in increased wire preheating, which tends to increase the deposit rate. Too much wire stick-out may also produce a ropy appearance in the weld bead. Too little stick-out will cause the wire to fuse to the nozzle tip, which decreases the life of the tip. As the amount of wire stick-out increases, it may become increasingly difficult to follow the weld seam, particularly with a small diameter wire. The tip should be either flush with the gas nozzle or recessed in the nozzle. See Fig. 12-8, *bottom.* An extended tip is seldom used and then for very low amperages.

The wire, in a near-plastic state between the tip and arc, tends to move (whip) around, describing a somewhat circular pattern. Decreasing the amount of wire stick-out and straightening the welding wire tend to decrease the amount of wire whip.

Welding Current

A wide range of current values can be used with each wire diameter. This permits welding various thicknesses of metal without having to change wire diameter. The correct current to use for a particular joint must often be determined by trial. The current selected should be high enough to secure the desired penetration without cold lapping (cold shuts) but low enough to avoid undercutting and burn-through. (See Tables 12-4, 12-5, 12-6, 12-7.)

The success of Mig welding is due to the concentration of a high-current density at the electrode tip. Whereas the arc stream of Mig is sharp and deeply penetrating, metallic arc (stick electrode) is soft and widespread. Consequently,

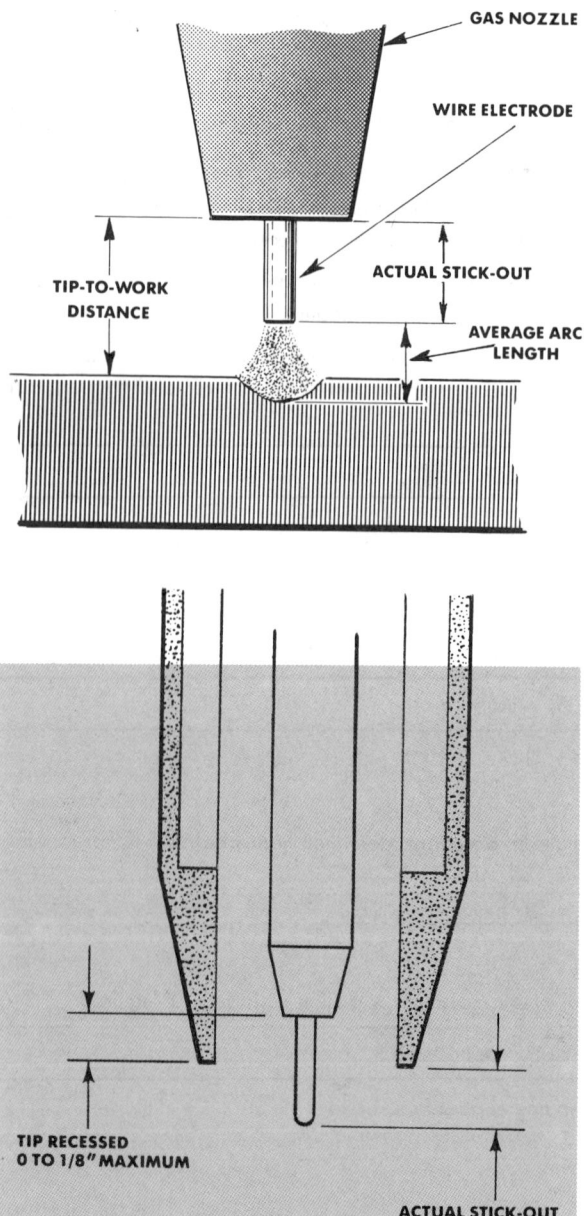

Fig. 12-8. Correct wire stick-out is important to achieve sound welds. (Hobart Brothers Co.)

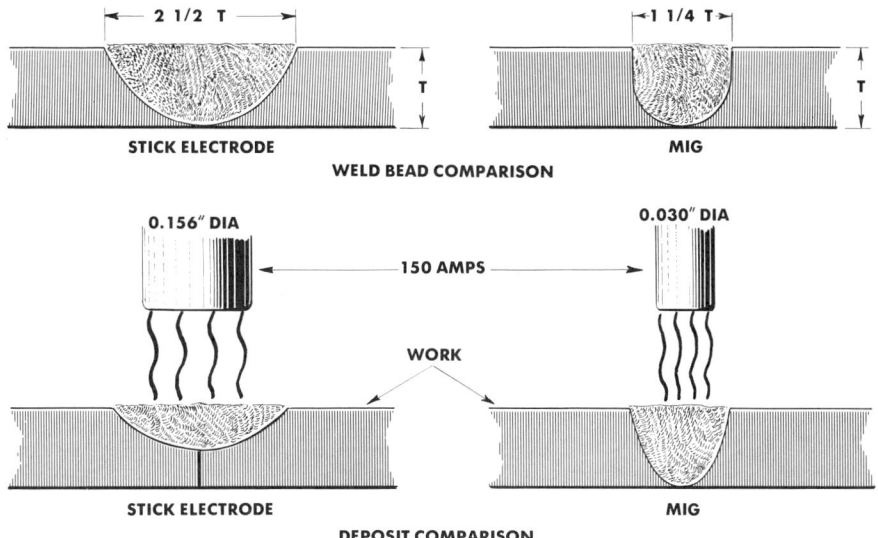

Fig. 12-9. The high-current density of Mig welding produces deep penetration and narrow beads.

the width-to-depth ratio of gas metal arc will be less than with stick electrode. See Fig. 12-9 for comparison.

Joint Edge Preparation and Weld Backing

Preparation of the edges of each member to be joined aids penetration and control of weld reinforcement. For Mig welding, beveling of the edges is usually desirable for butt joints thicker than 1/4" if complete root penetration is desired. For thinner sections, a square butt joint is best.

To a considerable extent the same conventional joint design recommended for other arc welding processes can be used for gas metal-arc welding. However, some joint modifications are often incorporated to compensate for the operating characteristics of gas metal-arc welding. Thus the arc in gas metal-arc welding is more penetrating and narrower than the arc in shielded metal-arc welding. Consequently, groove joints can have smaller root faces and root openings. Also, since the nozzle does not have to be placed within the groove a narrower included angle can be provided. By reducing the joint area less weld metal is required; this lowers material and labor costs.

In the Mig process, weld backing is helpful in obtaining a sound weld at the roots. Backing prevents molten metal from running through the joint being welded, especially when complete weld penetration is desired.

There are several types of material used for backing: steel and copper blocks, strips, and bars, carbon blocks, plastics, asbestos, and fire clay. Some of these serve to conduct heat away from the joint and also to form a mold or dam for the metal. The most commonly used backing for Mig welding is copper or steel.

Work Position and Welding Wire

The proper position of the welding torch and weldment is important. In Mig welding, the flat position is preferred for most joints, because this position improves the molten metal flow, bead contour, and gives better gas protection. However, on gaged material, it is sometimes necessary or advantageous to weld with the work inclined 10° to 20°. The welding is done in the downhill position. This has a tendency to flatten the bead and increase the travel speed.

The alignment of the welding wire in relation

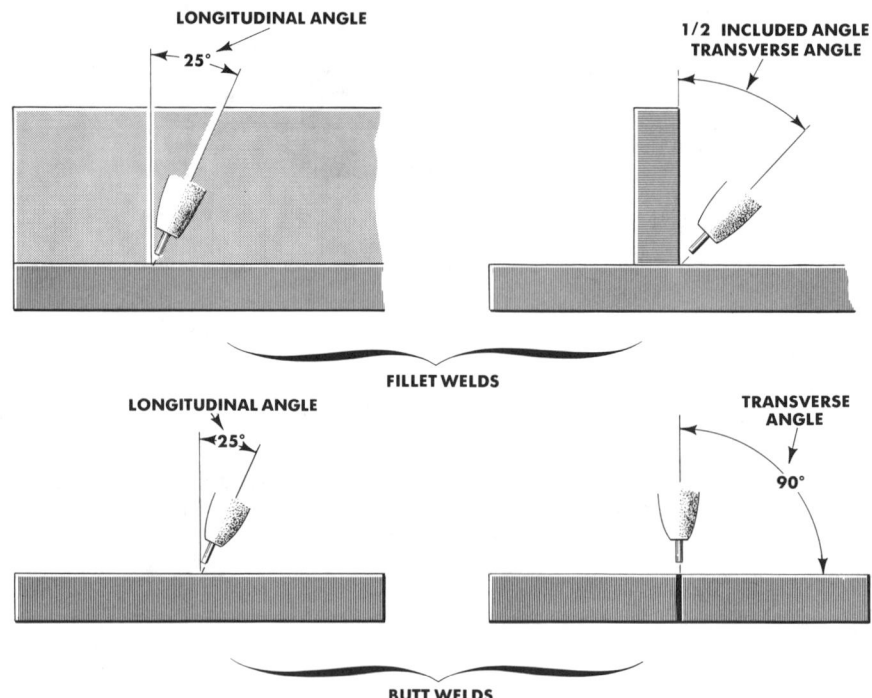

Fig. 12-10. Correct nozzle angle for Mig welding.

to the joint is very important. The welding wire should be on the center line of the joint for most butt joints, if the pieces to be joined are of equal thickness. If the pieces are unequal in thickness, the wire may be moved toward the thicker piece. The recommended position of the welding gun for fillet and butt welds is shown in Fig. 12-10.

Either a pulling or pushing technique may be used with little or no weaving motion. See Fig. 12-11. Some weaving is desirable for poorly fitted edge joints. The pulling or drag technique is usually best for light gage metals and the pushing technique for heavy materials. In the pulling technique (backhand) the gun points away from the direction of travel, whereas in the pushing motion (forehand) the gun points foreward in the direction of travel.

Generally the penetration of beads deposited with a pulling technique is greater than with a pushing technique. Furthermore, since the welder can see the weld crater easier in a pulling action he can produce high quality welds more consistently. On the other hand, forehand weld-

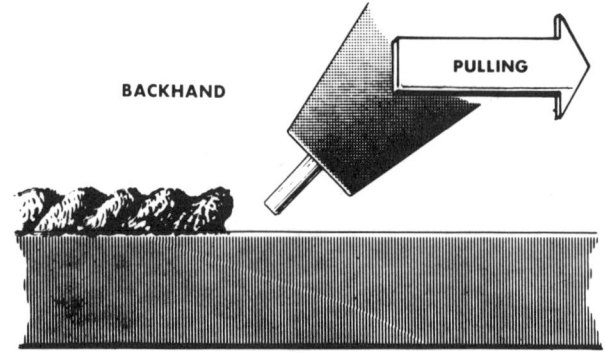
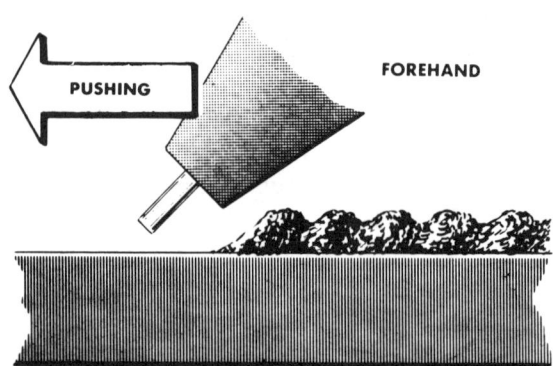

Fig. 12-11. Use either a pushing or a pulling technique in Mig welding.

ing permits the use of higher welding speeds and produces shallower and wider welds.

For welding circular seams, the wire should be shifted off-center, approximately $1/3$ of the work radius, as shown in Fig. 12-12. This will allow the metal to solidify by the time it reaches the top of the circle. A shift of more than $1/3$ of the work radius will cause the weld metal deposit to run ahead of the weld bead.

Preliminary Welding Checks for Mig Welding

Before starting to weld, it is always a good practice to check the following:

1. All electric power controls are in the OFF position.

2. All hose and cable connections from the gun to the feeder are in good condition, properly insulated, and connections have been correctly made and secured.

3. Nozzle for the diameter wire is correct.

4. Wire is properly threaded through gun.

5. Apertures of contact tube and nozzle are clean. (Blow out the gun occasionally, because sometimes it becomes loaded with dust and restricts proper wire feed and flow of protective gases.)

6. Wire speed and feed have been predetermined and adjusted on the feeder control.

7. Shielding gas and water coolant sources are on and adjusted for desired output.

8. Wire stick-out is correct.

9. Contact tip is in proper shape. (Tips eventually wear out, especially under high usage, and must be replaced for good welds.)

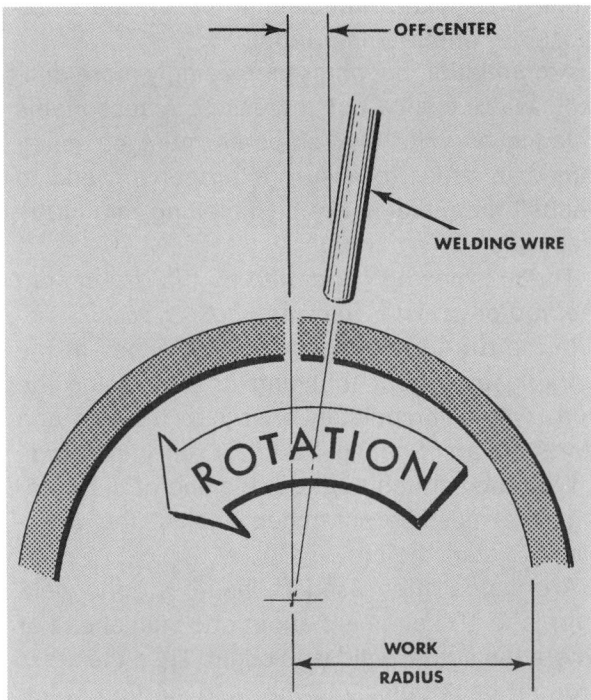

Fig. 12-12. Position of electrode wire in making a circular weld. (Hobart Brothers Co.)

WELDING PROCEDURE

In general, Mig welding procedure follows a definite sequence, regardless of the kind of welding done. Basically the following steps are involved:

1. Set the voltage, wire feed, and gas flow to the prescribed conditions for the required type of welding. During welding the wire speed rate may have to be varied to correct too much or too little heat input.

2. Adjust the wire for the proper amount of stick-out.

3. Start the arc and move the gun along the seam at a uniform speed, keeping the gun at the correct angle. If the arc is not started properly, the filler wire may stick to the work or actually freeze to the tip. Should this happen, shut off the machine and free the wire.

4. Move the gun along the seam with a pushing or pulling motion. As the gun is moved keep the wire at the leading edge of the puddle. Also be sure the wire is centered in the gas pattern to insure adequate shielding.

5. Release the trigger when reaching the end of the weld. Releasing the trigger stops the wire feed and interrupts the welding current. However, always keep the gun over the weld until the gas stops flowing in order to protect the puddle until it solidifies.

6. Shut down the welding unit when welding is completed, following this sequence:
   a. Turn OFF wire speed control.
   b. Shut OFF gas flow at cylinders.
   c. Squeeze the welding gun trigger to bleed the lines.
   d. Hang up the welding gun.
   e. Shut OFF welding machine.

During any welding operation certain welding condition may have to be changed. Some of the more specific welding variables with their required changes are shown in Table 12-3.

TABLE 12-3   CORRECTING WELDING VARIABLES.

CHANGE DESIRED	ACTION REQUIRED
1. Deeper penetration	Increase welding current, or decrease wire stick-out, or use smaller wire size
2. Shallower penetration	Decrease welding current, or increase wire stick-out, or use larger wire size
3. Larger bead	Increase welding current, or decrease travel speed, or increase wire stick-out
4. Smaller bead	Decrease welding current, or increase travel speed, or decrease wire stick-out
5. Flatter, wider bead	Increase arc voltage, or decrease wire stick-out
6. Faster deposition rate	Increase welding current, or increase wire stick-out, or use smaller wire size
7. Slower deposition rate	Decrease welding current, or decrease wire stick-out, or use larger wire size

Arc Starting

Starting an electrical arc for a welding process involves three major factors: electrical contact, arc voltage, and time. To assure good arc starts, it is necessary for the electrode wire to make good electrical contact with the work. The electrode must exert sufficient force on the workpiece to penetrate impurities.

Arc initiation becomes increasingly more difficult as wire stick-out increases. A reasonable balance of volts and amperes must be maintained in order to assure a proper arc and to deposit the metal at the best melting rate of the electrode.

The arc may be generated by the *run-in start method* or *scratch start method.*

In the run-in method the gun is aimed at the workpiece without touching it. Depressing the gun trigger immediately energizes the wire and starts the arc. See Fig. 12-13 for run-in method.

With the scratch method, the end of the welding wire must be scratched against the workpiece to start the arc.

A practice often used to insure a good weld start is to strike the arc about one inch ahead of where the actual weld is to begin. Then the arc is brought back quickly to the weld starting point.

Fig. 12-13. Start the arc by pressing gun trigger. (Hobart Brothers Co.)

Gas Metal-Arc Welding

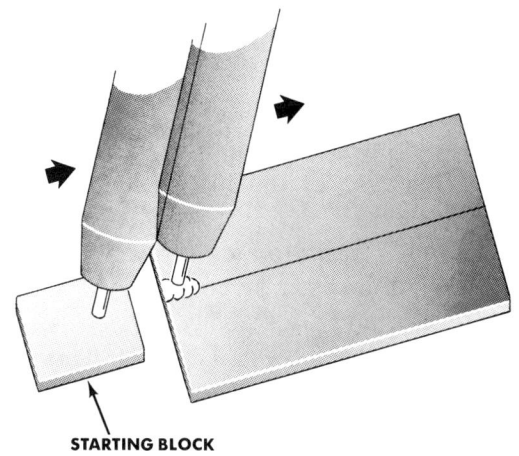

Fig. 12-14. Strike the arc ahead of the starting point of the weld.

leading edge of the puddle. See Fig. 12-15. If undercutting occurs, use a slight side-to-side weaving motion. See Figs. 12-16 and 12-17.

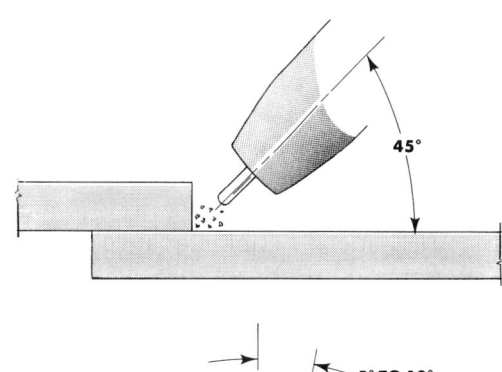

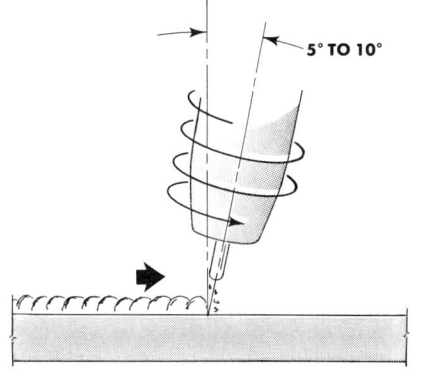

Fig. 12-16. Mig welding a lap joint using a pulling technique.

Another way is to strike the arc on a starting block outside the weld seam. See Fig. 12-14.

When finishing a weld a similar technique is used. The direction of travel is reversed and at the same time the speed increased. This technique helps to taper the width of the molten pool before breaking the arc. By following such a procedure, the chance of leaving a crater in the last deposited bead decreases.

Once the arc is started, the gun is held at the correct angle and moved at a uniform speed.

Remember that travel speed controls the weld bead size. If you want a smaller bead, increase the rate of travel. For a larger bead reduce the rate of travel. Keep the arc stream near the

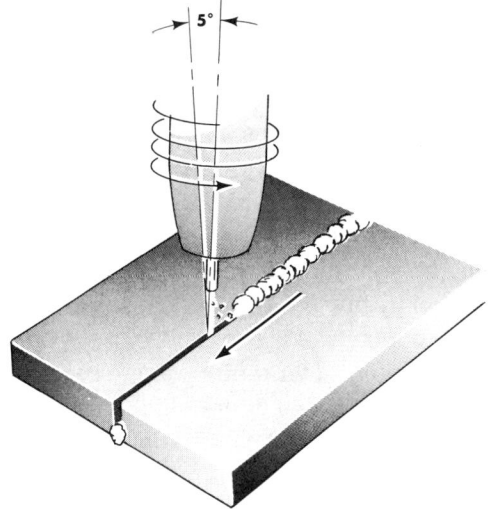

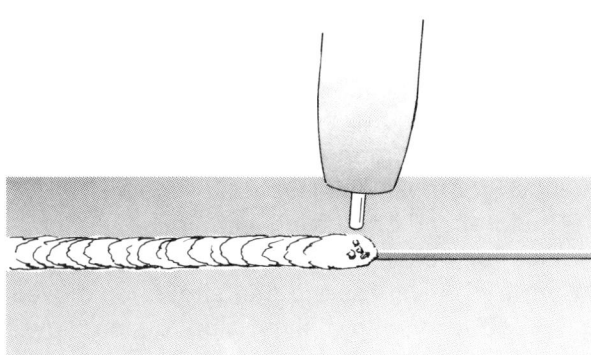

Fig. 12-15. Keep gun with wire in leading edge of puddle.

Fig. 12-17. Mig welding a butt joint using a pushing motion.

MIG WELDING COMMON METALS

Mig welding has become one of the most universally accepted processes for joining all types of metals. The ease with which sound welds can be produced by Mig welding has in many instances revolutionized welding practices in numerous industries. One of its particularly outstanding features is the ease with which production welding can be mechanized, thereby substantially reducing manufacturing costs.

Generally speaking, the same type of equipment and welding techniques apply to joining all metals. A few specific characteristics of welding several common metals are included in the following paragraphs.

Carbon Steels

Both the spray arc and the short arc produce excellent welds in carbon steels. For spray-arc welding a mixture containing 5 percent oxygen with argon is generally recommended. The addition of oxygen provides a more stable arc, minimizes undercutting and permits greater speeds.

Considerable amount of steel welding is done with a mixture of argon and CO_2. A straight CO_2 gas is sometimes used, especially for high-speed production welding.

For short arc welding of carbon and low-alloy steels, a 25 percent carbon dioxide and 75 percent argon mixture is preferred. The dioxide mixture improves arc stability and minimizes spatter.

Thin steel plates 0.035" to 1/8" in thickness may be butt-welded with square edges. Usually an opening of 1/16" or less is recommended. For wider openings the short arc is better since relatively large gaps are more easily bridged without excessive penetration. Plates 3/16" and 1/4" may be square butt-welded with a 1/16" to 3/32" root opening but usually two passes are necessary. For quality welds some beveling is desired. Plates 1/4" and greater require single or double-V grooves with 50 to 60 degree included angles.

Table 12-4 lists specific requirements for Mig welding carbon steels. On multi-pass welds the sequence of bead deposits is similar to metallic arc welding. See Unit 8.

Aluminum

The design for aluminum seams is similar to that for steel seams. However, narrower joint spacing and lower welding currents are recommended because of the higher fluidity of the metal.

Argon gas is preferred for Mig welding plates up to 1" in thickness since it provides better metal transfer and arc stability with less spatter. Sometimes in position welding of 1100 and 3003 aluminum the addition of a small amount of oxygen to the argon improves coalescence (flow of metals).

Welding aluminum with the *short arc* produces a colder arc than the spray type arc and thereby permits the weld puddle to solidify rapidly.

This action is especially advantageous in vertical, overhead, and horizontal welding, and in welding lighter materials. In vertical welding, a downhill technique is preferred.

Welding aluminum with the *spray arc* is especially suitable for thick sections. With spray arc more heat is produced to melt the wire and base metal. As a rule, vertical, horizontal and overhead welds are more difficult to make with the spray arc than with the short arc. See Table 12-5.

Stainless Steel

Copper back-up strips should be used to weld stainless steel up to 1/16" in thickness. Precaution must be taken to prevent air from reaching the underside of the weld while the puddle is solidifying since the oxygen and nitrogen will weaken the weld. To prevent air from contacting the underside of the weld, an argon back-up gas is often used.

Spray arc welding with 1/16" diameter wire and high current produces good welds. DCRP with a 1 or 2 percent argon-oxygen mixture is recommended for most stainless steel welding.

The forehand or pushing technique is generally used for welding stainless steel. On plates 1/4" or more in thickness the gun should be moved back and forth with a slight side-to-side movement. Thin materials are best welded with just a slight back-and-forth motion along the joint. See

TABLE 12-4. MIG WELDING—CARBON STEEL.

PLATE THICKNESS (inches)	JOINT AND EDGE PREPARATION	WIRE dia (inches)	GAS FLOW (cfh)	DCRP CURRENT (amps)		WIRE FEED (ipm)
0.035				55	16*	117
.047				65	17*	140
.063	Non-positioned	.030	10–15	85	17*	170
.078	fillet or lap			105	18*	225
.100			Mixture (75%A + 25% CO_2)	110	18*	225
1/8				130	19*	300
1/8	Butt (square edge)	1/16		280	—	165
3/16	Butt (square edge)	1/16		375	—	260
3/16	Fillet or lap	1/16		350	—	230
1/4	Double V butt (60° included angle, no nose)			375 (1st pass) 430 (2nd pass)	27	83 (1st) 95 (2nd)
5/16	Double V butt (60° included angle, no nose)		40–50	400 (1st pass) 420 (2nd pass)	28	87 (1st) 92 (2nd)
5/16	Non-positioned fillet		Mixture (95% argon + 5% O_2)	400		87
1/2	Double V butt (60° included angle, no nose)	3/32		400 (1st pass) 450 (2nd pass)		87 (1st) 100 (2nd)

*short arc
Linde Co.

TABLE 12-5. MIG WELDING—ALUMINUM (SPRAY-ARC).

PLATE THICKNESS (inches)	TYPE OF JOINT	WIRE dia (inches)	ARGON FLOW (cfh)	DCRP (amperes)	VOLTAGE (volts)	APPROXIMATE WIRE FEED (ipm)
0.040	Fillet or tight butt	0.030	30	40	15	240
.050	Fillet or tight butt	.030	15	50	15	290
.063	Fillet or tight butt	.030	15	60	15	340
.093	Fillet or tight butt	.030	15	90	15	410

Linde Co.

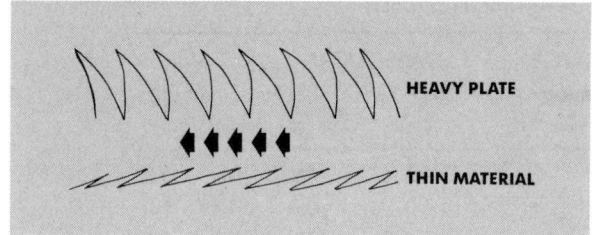

Fig. 12-18. Gun motion for Mig welding stainless steel.

Fig. 12-18. As a rule the short arc produces better welds on thin materials when overhead or vertical welding is required.

Tables 12-6 and 12-7 list specific requirements for Mig welding stainless steels.

Tubular Wire Welding (Flux-core)

Tubular-wire welding is a gas metal-arc welding process in which a continuous fluxed core wire instead of a solid wire serves as the electrode. The wire can be used on any automatic or semi-automatic Mig welding equipment and is used principally in combination with CO_2 as a shielding gas. The wire is frequently referred to by the manufacturer's trade name such as *Flux-cor* (Airco) and *FabCo* (Hobart).

The common AWS flux-cored wires are E-70T-1, E-70T-2, E-70T-3, and E-70T-4. The identity of the symbols is similar to those for solid wires except the letter T which designates a tubular wire.

TABLE 12-6. GENERAL WELDING CONDITIONS FOR SPRAY-ARC WELDING OF STAINLESS STEEL.

PLATE THICKNESS (inches)	JOINT AND EDGE PREPARATION	WIRE Dia	GAS FLOW	CURRENT DCRP (Amps)	WIRE FEED (ipm)	SPEED (imp)	WELDING PASSES
0.125	Square butt with backing	1/16	35	200–250	110–150	20	1
.250	Single V butt 60° inc. angle no nose	1/16	35	250–300	150–200	15	2
.375	Single V butt 60° inc. angle 1/16″ nose	1/16	(O_2–1)	275–325	225–250	20	2

Linde Co.

TABLE 12-7. GENERAL WELDING CONDITIONS FOR SHORT ARC WELDING OF STAINLESS STEEL.

PLATE THICKNESS	JOINT AND EDGE PREPARATION	WIRE dia (inches)	GAS FLOW (cfh)	CURRENT DCRP (amps)	VOLTAGE	WIRE FEED (ipm)	WELDING SPEED (ipm)	PASSES
0.063	Non-positioned fillet or lap	0.030	15–20	85	15	184	18	1
.063	Butt (square edge)	.030	O_2–2	85	15	184	20	1
.078	Non-positioned fillet or lap	.030	O_2–2	90	15	192	14	1
.078	Butt (square edge)	.030	O_2–2	90	15	192	12	1
.093	Non-positioned fillet or lap	.030	O_2–2	105	17	232	15	1
.125	Non-positioned fillet or lap	.030	O_2–2	125	17	280	16	1

*Voltage values are for C-25 gas or O_2-2 gas. For 90% HE—10% C-25, voltage will be 6 to 7 volts higher
Linde Co.

TABLE 12-8. FLUX-CORED ARC WELDING CONDITIONS—DCRP.

	MATERIAL THICKNESS (inches)[1]	CURRENT DCRP (amps)	ARC VOLTAGE	WIRE FEED ipm	SHIELDING GAS FLOW CFH[2]	TRAVEL SPEED ipm	NO. OF PASSES	WIRE STICK-OUT (inches)
flux-cored	1/8	300–350	24–26	100–120	35–40	25–30	1	3/4 to 1 1/2
arc welding	3/16	350–400	24–28	120–150	35–40	25–35	1	3/4 to 1 1/2
of steel	1/4	350–400	24–28	120–150	35–40	20–30	1	3/4 to 1 1/2
using 3/32″	3/8	475–500	28–30	180–210	35–40	15–20	1	3/4 to 1 1/2
diameter electrode	1/2	400–450	25–28	150–170	35–40	18–20	2–3	3/4 to 1 1/2
wire size	5/8	400–450	25–28	150–170	35–40	14–18	2–3	3/4 to 1 1/2
flat and horizontal positions	3/4	400–450	25–28	150–170	35–40	14–18	5–6	3/4 to 1 1/2

[1] For groove and fillet welds. Material thickness also indicates fillet size. Use V groove for 1/4 inch and thicker. Double V for 1/2 inch and thicker.
[2] Welding grade CO_2.

The flux ingredients in the wire include ionizers to stabilize the arc, deoxidizers to purge the deposit of gas and slag, and other metals to produce high strength, ductility and toughness in weld deposits. The flux generates a gas shield, which is augmented by the regular CO_2 shield, and a slag blanket that retards the cooling rate and protects the weld deposit as it solidifies.

Tubular wire is designed for high-current densities and deposition rates which, when combined with high-duty cycles, result in sharply increased production speeds. It is especially intended for application in large fillet single and multi-pass welds in either a horizontal or flat position using DCRP current. Because of its deep penetrating qualities into the weld root, tubular wire fillet welds of smaller leg size will have the same strength as stick fillet welds of larger size. For instance, double-welded butt joints up to 1/2″ thick can be welded without edge preparation.

Flux-core wire welding is similar to other Mig welding processes. See Table 12-8.

SELF QUIZ

Correct answers are listed in the back of the book.

Multiple Choice

Circle the letter which represents the correct answer.

1. Which of these factors is *not* associated with Mig welding?
    a. A special flux must be used over the weld seam
    b. wire feed and gas flow are automatic
    c. the heat-affected zone is narrower because of faster travel speed
    d. thin as well as thick materials can be welded
2. DCSP is impractical for the Mig process, because
    a. too much heat is concentrated at the puddle
    b. it produces too much penetration
    c. weld penetration is wide and shallow
    d. weld beads have uneven ripples

3. In DCRP the metal transfer from the electrode to the weld puddle is in the form of
   a. erratic globules
   b. fine spray
   c. fast but irregular spray
   d. oxide-free spray
4. The use of AC current is not recommended for Mig welding, because
   a. penetration is too shallow
   b. the danger of burn thru is too great
   c. welding speed is too slow
   d. wire burn-offs are unequal
5. Spray type transfer is not too practical for welding
   a. light gage metals
   b. heavy metal plates
   c. nonferrous metals
   d. joints with irregular root openings
6. One of the characteristics of the short arc is that
   a. it produces deep penetration
   b. it makes a more even metal spray possible
   c. it produces shallow penetration
   d. it is more effective for out-of-position welding
7. The current most effective for Mig welding is
   a. DCRP
   b. DCSP
   c. ACHF
   d. AC
8. A characteristic of a constant potential welding machine is that
   a. it produces a varying current over a wide range of welding situations
   b. it maintains the same voltage regardless of the amount of current drawn
   c. the wire feed must be adjusted to narrow limits
   d. if the distance from nozzle to work increases, the arc length increases
9. In a constant potential power supply unit
   a. the current automatically increases when the wire electrode feeds faster
   b. there is no increase in current when the arc length becomes shorter than the preselected value
   c. wire speed remains constant regardless of the change in arc length
   d. the voltage and wire feed increase if the arc becomes too long
10. The pull type gun for Mig welding is designed
    a. for large diameter wires
    b. to weld ferrous metals
    c. to weld with hard wires
    d. to handle small diameter wires
11. Which one of these mixtures is best for Mig welding low carbon steel?
    a. helium-oxygen
    b. argon-oxygen
    c. argon-helium
    d. CO_2-helium
12. Too little wire stick-out in the Mig gun will
    a. melt the wire too fast
    b. produce too much spatter
    c. cause wire to stick to the nozzle tip
    d. cause excessive wire whip

True and False

Circle the letter T if the statement is true or the letter F if it is false.

13. T F Mig welding produces deeper and narrower beads than the shielded metal arc.
14. T F For a specific voltage setting, a high wire feed will result in a long arc.
15. T F Wire stick-out influences the welding current, since it changes the preheating in the wire.
16. T F In a constant potential power supply unit there is an automatic adjustment of current when the wire is fed faster.
17. T F A higher amperage can be obtained by increasing the wire feed.
18. T F Burn thru can be reduced by using a faster speed feed.
19. T F For proper fusion the arc should be kept at the leading edge of the puddle.
20. T F Most joints designed for gas metal arc have a narrower beveled angle than the ones designed for shielded metal arc.
21. T F Micro type wires are used for the short arc.

22. T F The short arc produces a relatively cool weld puddle with less danger of burn thru.

Fill-In

Supply the missing word or words where blanks appear.

23. On light gage metals the gun is _____ along the weld seam.
24. On heavy metals the gun should point _____ in the direction of travel.
25. To correct too much or too little heat input the _____ may have to be adjusted.
26. The two methods which can be used to start the arc are _____ and _____.
27. The arc should be started about _____ in front where the weld is to begin.
28. To avoid leaving a crater, you should _____ the direction of travel when reaching the end of the weld and increase the _____ of the gun.

WELDING ASSIGNMENT

I. Depositing Surface Beads with Mig

1. Secure several pieces of low carbon steel and draw a series of straight lines on the surfaces. Ask your instructor what thickness of metal you are to use, since this has some bearing on the available wire electrode.
2. Make sure the wire stick-out is correct and the gun tip is recessed in the gas nozzle.
3. Set the required controls—wire feed, gas flow, voltage.
4. Practice starting and stopping the arc on one plate. Next run a few short lengths of beads. Continue to do this until you can readily start the arc without the wire sticking to the workpiece.
5. On the same plate, experiment with different lengths of wire stick-out. Begin with a very short wire stick-out, one that is less than the normally recommended length for the diameter wire which you are using. Then try depositing beads with a long wire stick-out. Note how these wire stick-outs affect the deposit rate, bead penetration, ease of maintaining the arc, appearance of the weld bead and amount of wire whip.
6. Try another experiment by observing the effects different wire feeds have on weld deposits. First set your wire feed control on the high side—9 or 10. This setting will give you a high amperage and result in more metal deposition than you need. Here the weld puddle is likely to be excessively large and the wire may even jab into your work. Next set the feed speed control to the low position—2 or 3. The amperage now will be low, giving you an unstable arc with irregular weld metal deposits, and the wire may stick to the nozzle tip.
7. After you have found the best wire feed speed, run a series of continuous straight beads on the drawn lines of your practice plate. Use both a pushing and pulling technique with no weaving motion. Simply slant the gun in the right direction and move it forward in a steady straight line.
8. Continue to run straight beads until you have mastered the knack of depositing beads of uniform width and height.

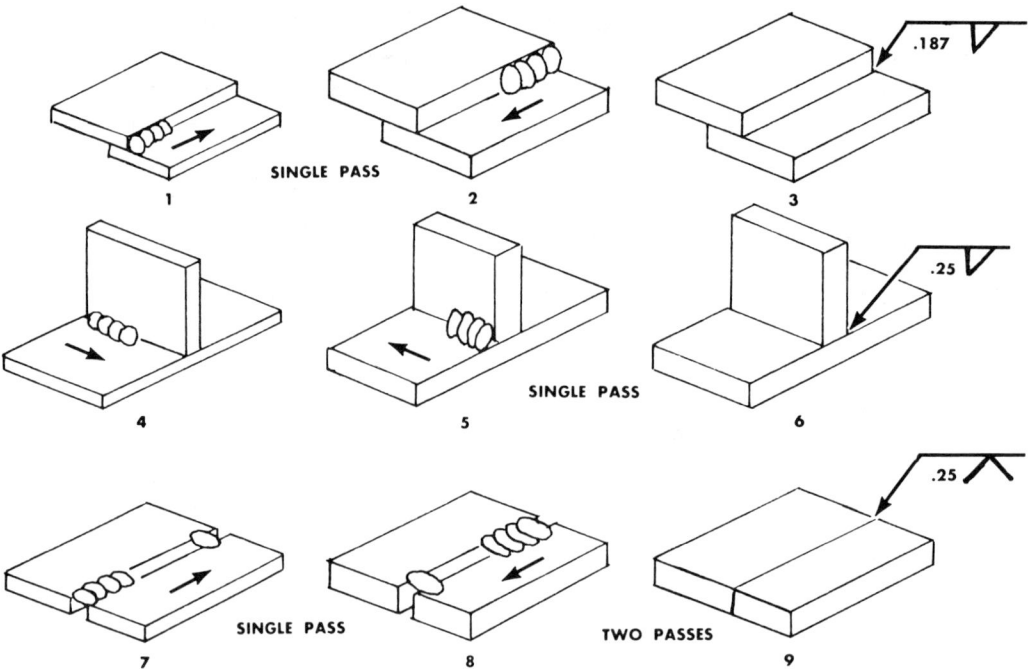

Fig. 12-19. Practice Mig welding sequence in flat position.

II. Welding Joints with Mig in a Flat Position

1. Practice welding lap, T and butt joints in the flat position. See Fig. 12-19. Check with your instructor about the metal thicknesses you are to use.

2. On lap and T joints hold the gun so the nozzle bisects the joint angle and lean it about 15-25 degrees in the direction of travel. Practice using both a pulling and pushing technique without any weaving motion.

3. On many jobs the size of a fillet weld is specified. Therefore, you need to practice depositing beads to meet size requirements. Travel speed will normally control weld bead size. Thus if you want to make a smaller bead, increase your travel speed. If you want a larger bead, reduce the travel speed.

4. When welding butt joints, practice using a single pass on square edge joints and multiple pass on beveled joints. Use a small crescent or rotary motion on the final pass.

5. Test each weld by bending it in a vise. Analyze your welds by using the Weld Analysis Check List at the end of this unit.

III. Welding Joints with Mig in a Vertical Position

1. Secure several pieces of low carbon steel and tack weld them to form lap, T and butt joints. Check with your instructor about the metal thickness you are to use.

2. Weld these joints in a vertical position. Practice both uphill and downhill welding techniques.

3. Follow the practice welding sequence shown in Fig. 12-20. Make bead sizes as specified.

4. Test each weld.

IV. Welding Joints with Mig in a Horizontal Position

1. Ask your instructor what thicknesses of low carbon steel you are to weld.

2. Practice welding the joints as shown in Fig. 12-21 in a horizontal position. Follow the weld sequence as indicated and incorporate the specifications given.

3. Test each completed weld by using the

Gas Metal-Arc Welding **133**

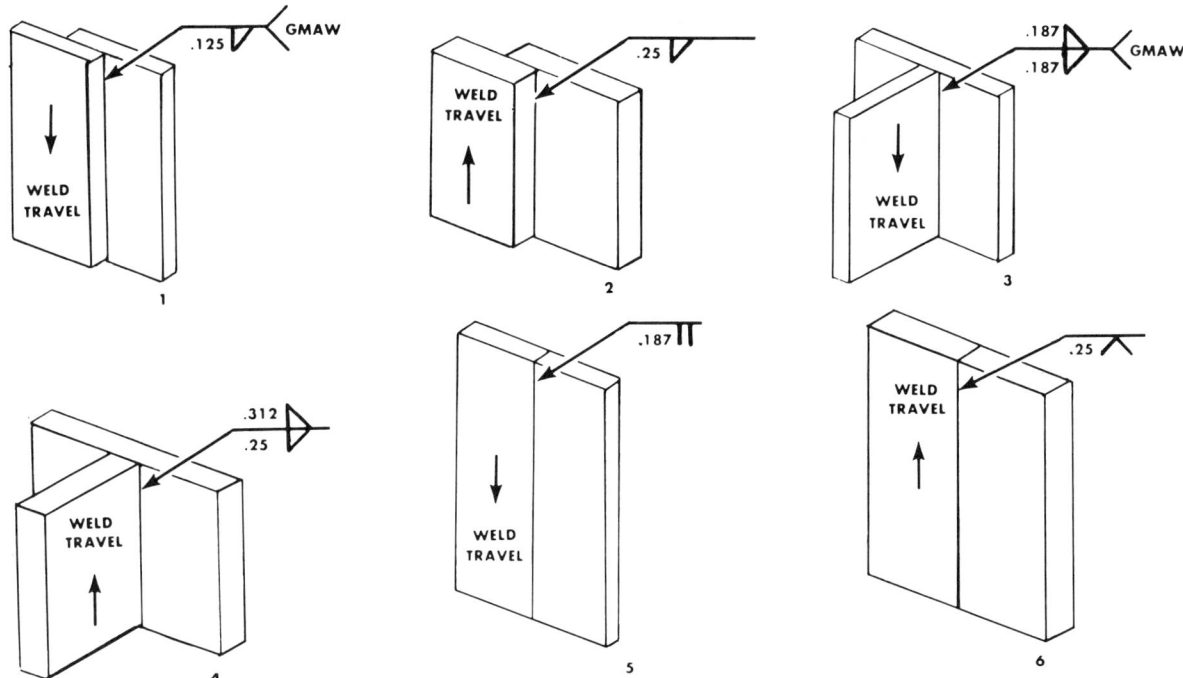

Fig. 12-20. Practice vertical Mig welding sequence.

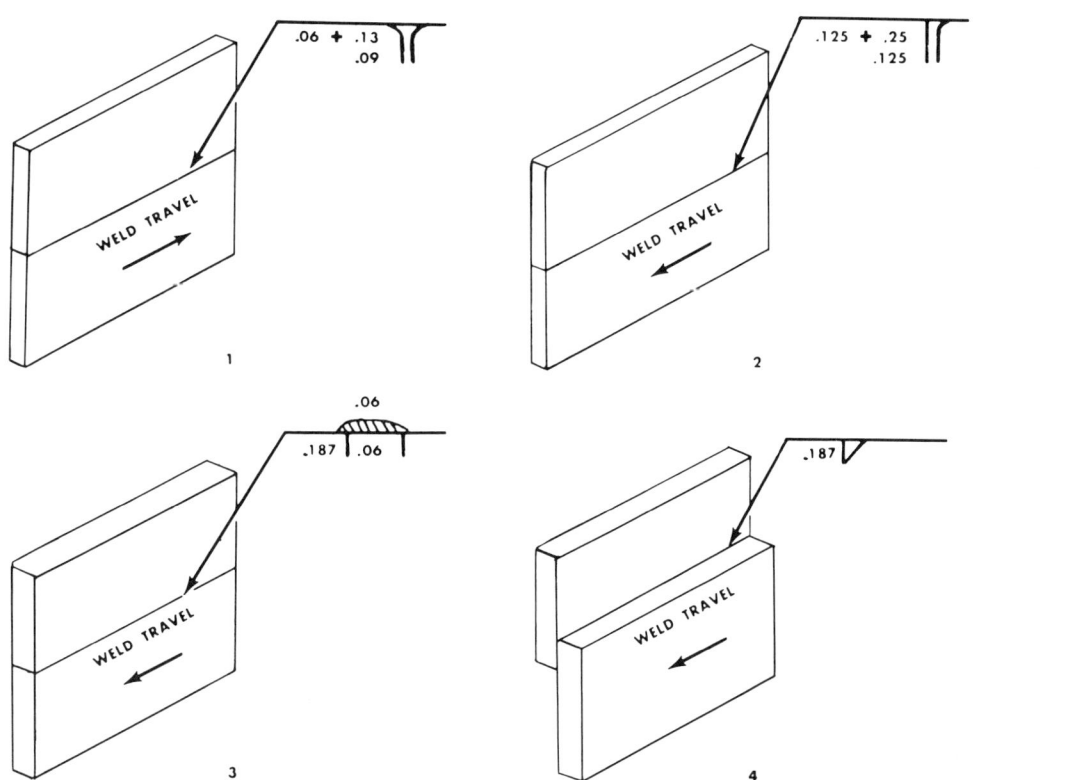

Fig. 12-21. Practice weld sequence for horizontal Mig welding.

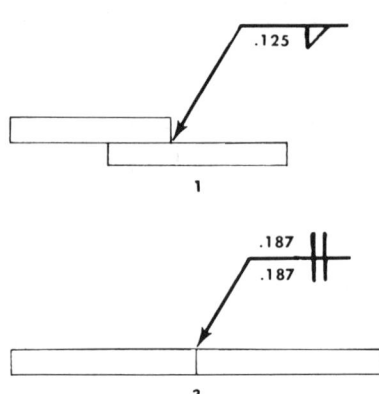
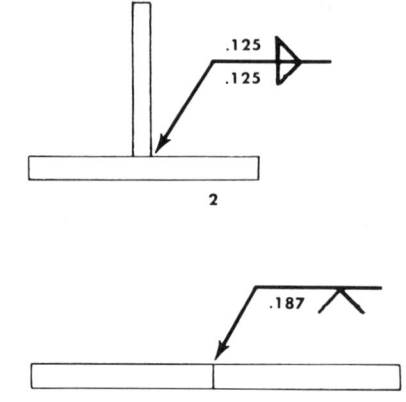

Fig. 12-22. Practice sequence for overhead Mig welding.

bend method. Evaluate your results by referring to the Weld Analysis Check List at the end of this unit.

V. Welding Joints with Mig in an Overhead Position

1. Practice welding lap, T and butt joints in the overhead position. Select suitable thicknesses of low carbon steel plates. Use both pushing and pulling welding techniques.
2. Follow welding sequence and weld specifications as shown in Fig. 12-22.
3. Test each completed weld and analyze the test results.

VI. Welding Stainless Steel with Mig

1. Ask your instructor about the type and thickness of stainless steel which you are to weld.
2. Practice welding edge, lap, T and butt joints in the flat, vertical, horizontal and overhead positions. See Fig. 12-23 for the weld sequence you are to follow. Always be sure the arc is started properly; otherwise the wire may stick to the work. When this happens the wire will break or bend at the end of the contact tube and freeze to the tube. If this should occur, stop the machine and free the wire.
3. Test each weld and analyze the test results.

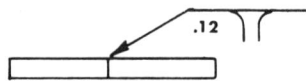
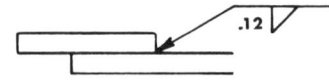
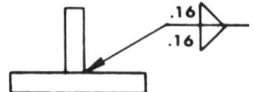

Fig. 12-23. Weld sequence for stainless steel Mig welding.

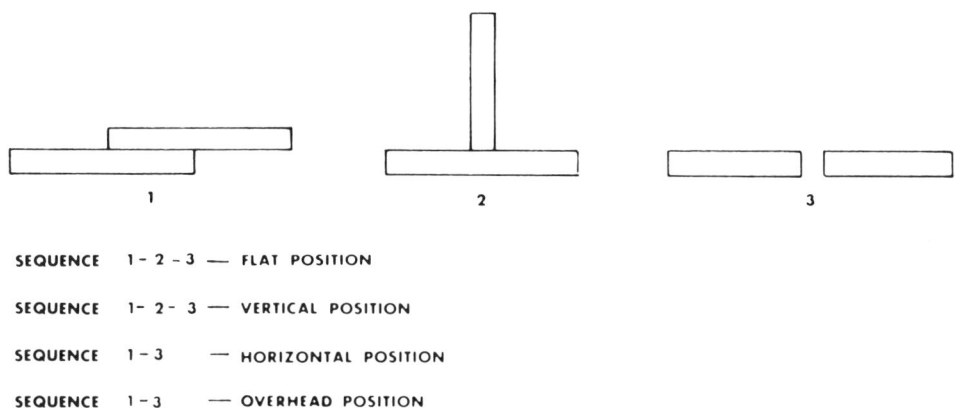

```
SEQUENCE   1-2-3   — FLAT POSITION
SEQUENCE   1-2-3   — VERTICAL POSITION
SEQUENCE   1-3     — HORIZONTAL POSITION
SEQUENCE   1-3     — OVERHEAD POSITION
```

Fig. 12-24. Practice sequence for Mig aluminum welding.

VII. Welding Aluminum with Mig

1. Ask your instructor what aluminums you are to weld. Then make the necessary decisions concerning wire feed, gas flow, voltage as well as any welding procedure most suitable for the metal to be welded.

2. Practice welding lap, T and butt joints in flat, vertical, horizontal and overhead positions. Follow the welding sequence shown in Fig. 12-24.

3. Test all completed welds.

Weld Analysis Check List	Yes	No
1. Bead widths are right size		
2. Beads have uniform ripples		
3. Weld beads are too flat		
4. Weld beads are too high		
5. Weld penetration is insufficient		
6. Weld penetration is excessive		
7. Cold laps on surface		
8. Weld has surface porosity		
9. Weld has subsurface porosity		
10. Weld has crater cracks		
11. Weld has burn thru		
12. End crater is filled		
13. Weld passed bend test without cracking		

UNIT 13
Pipe Welding

Welding is the easiest and simplest method of joining sections of pipe together since it eliminates complicated threaded joint designs, permits free flow of liquids, and reduces installation costs.

Welding is also considered a practical, effective, and cost-cutting technique in joining low-pressure piping systems for refrigeration, air-conditioning or heating applications.

Although some pipe sizes are occasionally welded with oxy-acetylene, most pipe welding is done with the shielded metal arc. However, much pipe is now welded with Mig, either manually, semi-automatically or automatically.

Pipe welding is recognized as a specialty in itself. See Fig. 13-1. Although many of the skills and practices required in pipe welding are similar to those required in other types of welding, pipe welders usually must develop certain techniques that are characteristic of pipe welding

Fig. 13-1. Welding is used extensively in joining pipe. (Hobart Brothers Co.)

alone. Furthermore, since public health, environmental restrictions and safety are involved, especially in welding cross-country transmission pipelines and high-pressure lines that are to convey steam, oil, air, and corrosive materials, pipe welders always have to pass certain tests to be certified.

Roll and Position Welding

Pipe welding in the field is done in several ways. In one method two or more sections are lined up and tack welded. Special pipe clamps, as shown in Fig. 13-2, are used to hold the pipe in alignment until they are tacked. The weld is then completed in the flat position while helpers rotate the pipe. This operation is called *roll welding*.

The *stove pipe* or position method consists of lining up each section, length by length, and welding each joint while the pipe remains stationary. Since the pipe is not revolved, the welding has to be done in various positions—flat, horizontal, vertical, and overhead.

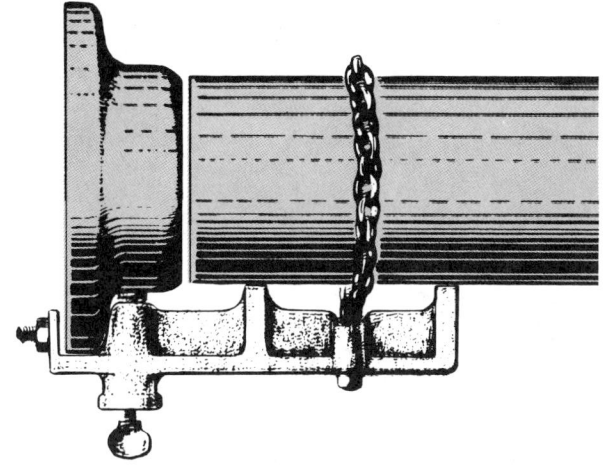

Fig. 13-2. Fast-action clamps hold the pipe in alignment while tack welds are made.

As mentioned, most pipe welding is done either with the shielded metal arc or gas metal arc (Mig). Gas tungsten arc (Tig) is occasionally used in shop welding to weld small diameter pipes. See Fig. 13-3. However, Tig is also used in certain pipe jobs to lay the root bead in large diameter pipes.

Fig. 13-3. Tig welding is often used to join small diameter pipe. (Hobart Brothers Co.)

138 Arc Welding

Fig. 13-4. A typical pipe beveling unit. (DND Corp.)

The advantage of Mig over stick welding is that no slag inclusions occur in the pipe weld. Since no slag has to be removed, less welding time is required. Furthermore, better welds can be made because the gas shield protects the weld area from atmospheric contamination.

Insofar as welding techniques and procedures are concerned, there is no significant difference between the shielded metal arc and the gas metal arc processes. Therefore, the general pipe welding techniques, a description of which follows, apply to both stick and Mig welding.

Pipe Joint Preparation

For welding most pipes, a single-V joint is used. The beveling is usually done with a regular oxy-acetylene beveling machine, as shown in Fig. 13-4.

The standard joint specifications for thick-wall and thin-wall pipe are shown in Fig. 13-5. Generally, pipes having wall thicknesses of 1/8" to 5/16" are classified as thin-wall pipe and pipes with wall thicknesses over 5/16" as thick-wall pipe regardless of diameter. Notice in Fig. 13-5 how

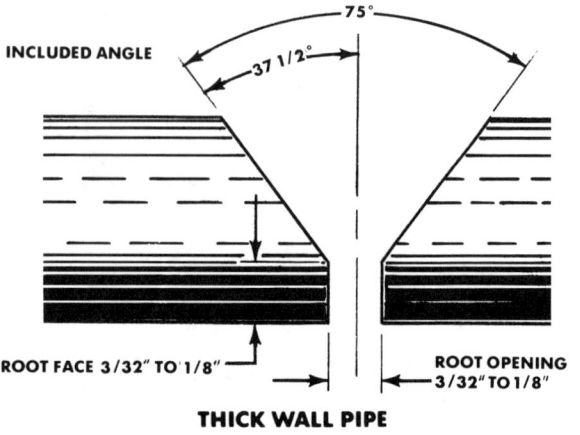
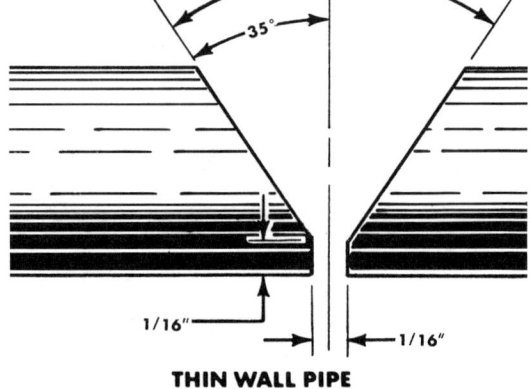

Fig. 13-5. Standard bevel specifications for thin and thick-wall pipe.

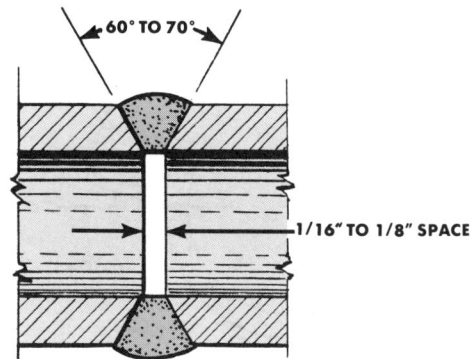

Fig. 13-6. Small diameter pipes with thin walls are usually butt welded.

the included angle, root face, and root opening will vary for both thin-wall and thick-wall pipe.

Small diameter pipes with wall thicknesses of less than 1/8" are normally welded without any edge preparation. The ends are simply butted together with a small separation to insure complete fusion. See Fig. 13-6. This classification of pipe is frequently welded by the Mig or the Tig methods.

Occasionally, liners or backing rings are fitted into the pipe before welding. These rings assist the welder to secure penetration without burning through the surface as well as to prevent spatter and slag from entering the pipe at the joint. Backing rings are also useful to keep pipes in alignment and stop metal icicles from forming on the inside of the joint, Fig. 13-7.

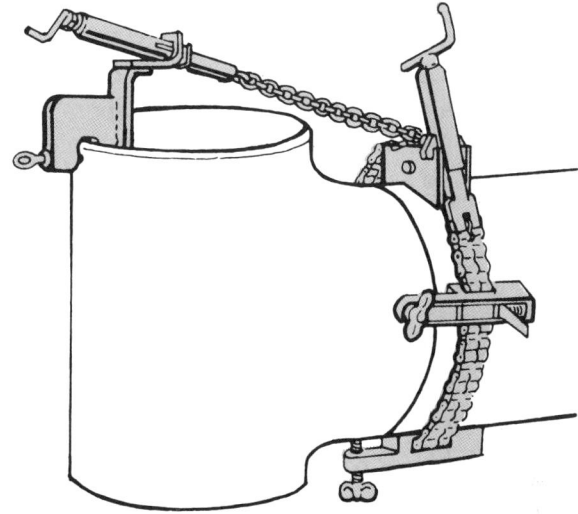

Fig. 13-8. Special clamps hold the pipe in alignment while tack welds are made. (CRC-Crose International, Inc.)

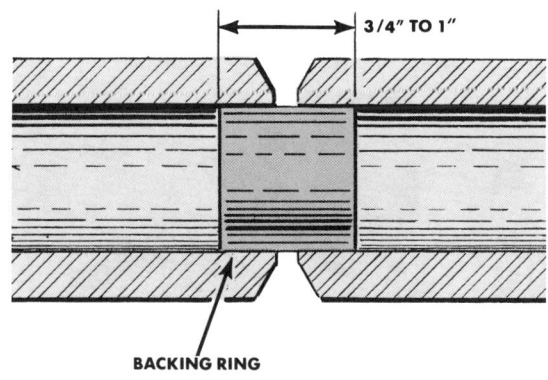

Fig. 13-7. Backing rings are sometimes fitted in small diameter pipes before welding. (Hobart Brothers Co.)

Tack Welding

Before welding, pipes are properly aligned and then tack welded. Special line-up clamps are used to insure correct alignment. See Fig. 13-8.

To maintain the required spacing between pipe sections a spacing tool is necessary. A wire of correct diameter placed between the pipes will provide the proper spacing.

For most pipe welding, four tack welds are made. These tack welds are evenly spaced around the pipe and approximately 3/4" long. Tack welds should penetrate to the root of the groove since they become part of the root bead.

To make a tack weld, the electrode is inclined

140 Arc Welding

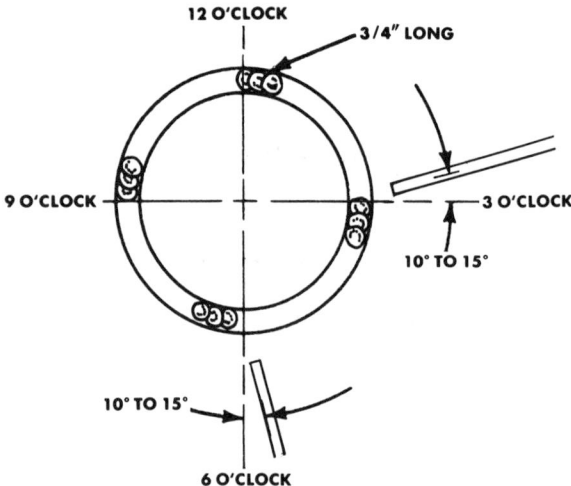

Fig. 13-9. Tack welding a pipe.

10 to 15 degrees as shown in Fig. 13-9. The arc is struck in the joint slightly ahead of where the weld is to be made. Then the arc is quickly lengthened so as to stabilize it and give it time to form a protective gas shield. Now the electrode is pushed into the joint with a light pressure and a sliding motion started in the groove. If the electrode has a tendency to stick, it should be wiggled slightly but kept buried in the groove. When the tack weld is completed, the electrode is pulled away. This procedure produces a strong and fully penetrating tack weld.

Welding Thin-Wall Pipe (Downhill Technique)

Most thin-wall pipe ($1/8''$–$5/16''$) is welded by using the downhill technique. It is usually preferred for welding cross-country pipelines because it is faster.

The weld is started at the top or 12 o'clock position and carried downward to the bottom or 6 o'clock point of the pipe. After the 6 o'clock position is reached, the same procedure is followed on the opposite side. See Fig. 13-10.

One of the problems in downhill welding is

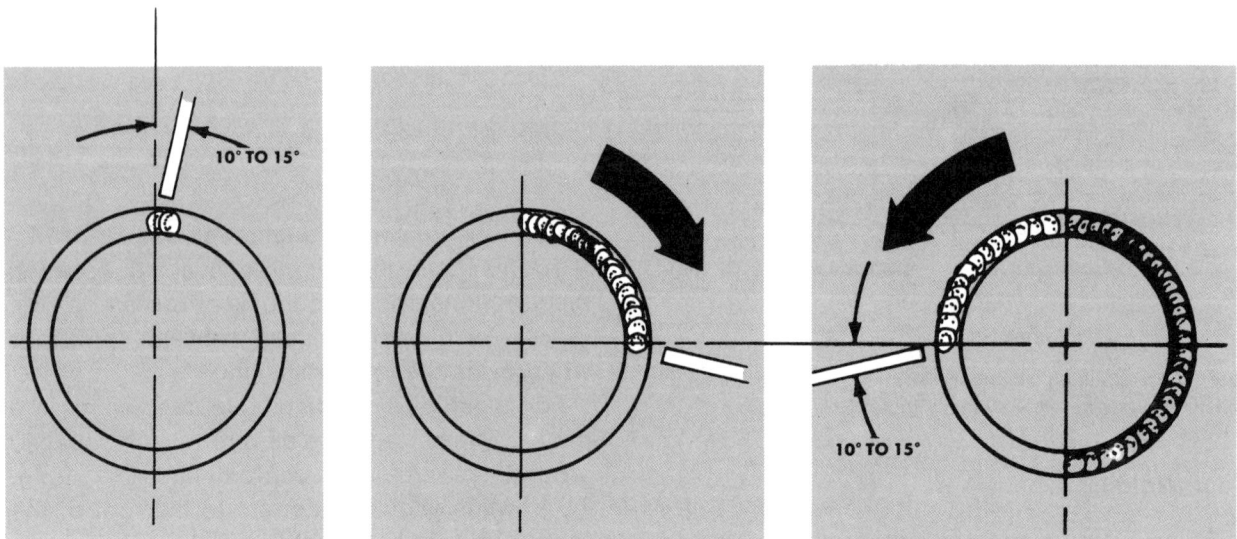

Fig. 13-10. Position of the rod for downhill welding.

controlling the heat input. This is particularly true in welding small diameter pipe where the heat does not dissipate fast enough and excessive heat builds up in the weld zone. Generally, heat input can be regulated by using a smaller diameter electrode and reducing the current setting.

Another problem in downhill welding is proper control of the puddle. The molten metal tends to flow downward in the same direction the arc is moving. If this is not controlled, penetration cannot be achieved and slag becomes entrapped in the molten metal, thereby producing slag inclusions in the weld. Slag inclusion, of course, is no problem in Mig welding. Control of metal flow is accomplished by keeping the arc ahead of the puddle. This can be done by using a fast travel speed and a high-current setting.

Welding procedure. After the pipes are securely tacked, a root bead is made completely around the joint. The electrode is held in approximately the same position as in making tack welds. The arc is struck slightly ahead of the weld to preheat the area where the weld bead is to be started. After the arc is stabilized, the electrode is lowered into the root opening and dragged along the edges of the groove. If the electrode has a tendency to stick and fails to glide smoothly because of the built-up heat, a slight side-to-side oscillating motion usually corrects the problem.

A properly deposited root bead should penetrate to the root and leave a solid bead below the surface with a slight crown not to exceed $1/16''$. See Fig. 13-11. Some undercutting may occur on the faces of the groove but this is not objectionable since this defect will be eliminated by successive passes.

There will be times when the root opening will vary due to poor fit-up. If the root opening is narrow, the speed of travel and electrode angle should be reduced. Where a widened root opening exists the travel should be increased.

The success of a pipe weld depends on the correct penetration of the root bead because it forms the base upon which the successive layers are made. Without full penetration the final weld joint will not be sound.

Starting and stopping. Since a welder has to start and stop a weld many times, due to changing electrodes or weld position, careful attention must be given to tying the ends of the weld together. To restart a weld the arc is struck about $1/2''$ back of the bead and then moved forward with a long arc. As soon as the arc is stabilized, the electrode is momentarily buried in the crater of the last beads to generate a pool of molten metal. The electrode is then raised slightly and the weld continued.

When a weld approaches the end and must be tied into the last deposited bead, the electrode is moved up the sloping sides of the previous bead,

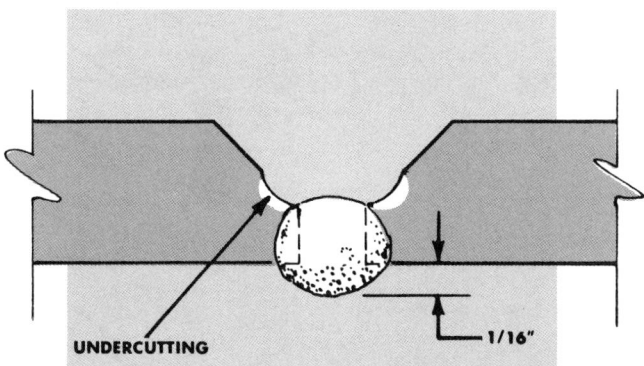

Fig. 13-11. Correct root bead penetration.

and after the molten puddle blends smoothly between the two beads, the direction of travel is briefly reversed. The arc is then withdrawn quickly by flicking the electrode downward and away from the center.

Successive passes. Upon completion of the root bead, additional layers of weld are deposited. The number of passes depends on the thickness of the pipe. Usually the subsequent layers consist of one hot pass, one or more filler passes and a final *cover* or *cap pass.* The specific function of the *hot pass* is to burn out the remaining particles of slag that may exist in the groove and to achieve a more complete fusion between the base metal and the root bead. This pass usually consists of only a light bead deposit and is made with a whipping motion as shown in Fig. 13-12. The electrode is moved down and up a distance of approximately 1½ electrode diameters, pausing momentarily at the top of the up motion. The whipping action permits better control of the weld puddle and forces the metal to flow into the undercuts. The hot pass is made with the same diameter electrode used for the root bead but with slightly higher current.

The intermediate or filler passes are deposited with larger diameter electrodes and are intended to fill the weld joint. After each pass is completed, the slag must be entirely removed if stick welding is used. Each layer should start and end at a different point to insure a uniform strong weld bond. Thus if the first pass is made at the 12 o'clock position, the next pass should begin about one inch below that spot.

Filler passes are normally made with a slant motion. See Fig. 13-13, top. The side-to-side weave should travel downward a distance of about one electrode diameter per stroke. The electrode should pause at the end of each stroke to insure good fusion at each edge of the weld. As the electrode reaches the bottom of the weld or is in the 6 o'clock position, a semi-circle or

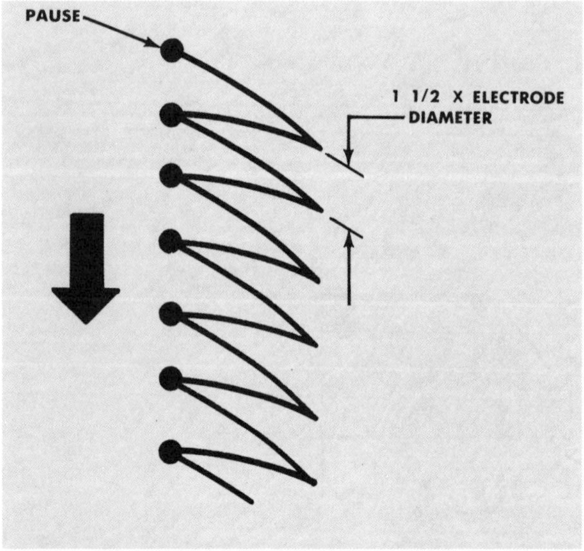

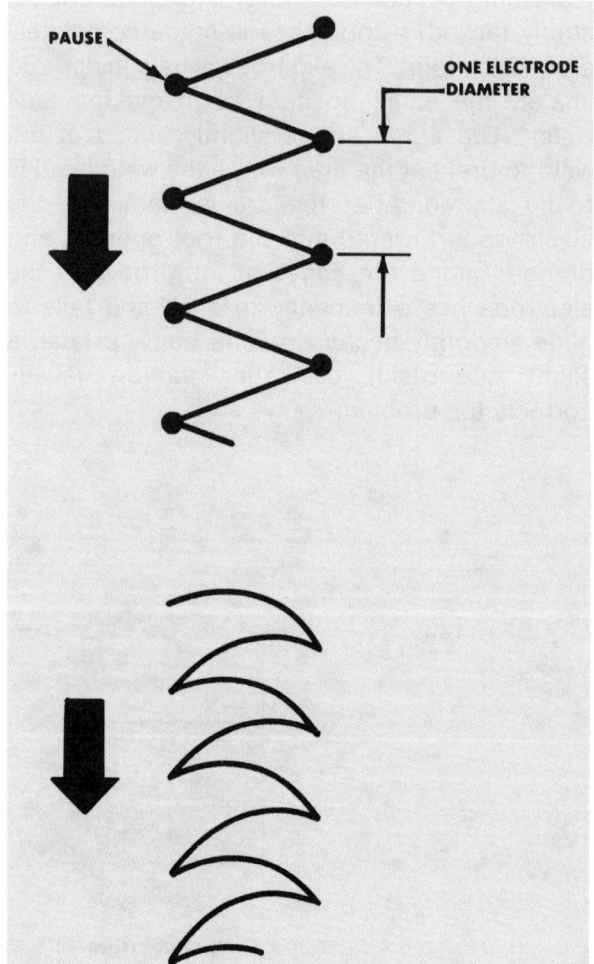

Fig. 13-12. Whipping motion used in making a hot pass.

Fig. 13-13. Motion used in filler passes.

horseshoe weave is often used. See Fig. 13-13, bottom. This motion permits better control of the puddle since it reduces the fluidity of the molten metal.

The final cover or cap pass is intended to provide maximum reinforcement to the weld joint and at the same time give the weld a neat appearance. The cover pass should have a slight crown extending about $1/16''$ above the surface of the pipe. Either a slant or semi-circular motion can be used. However, the weave must be wide enough to cover the entire weld joint.

Welding Heavy-Wall Pipe (Uphill Technique)

Uphill welding is basically intended to join heavy-wall pipe. The welding progresses upward on one side of the pipe and then upward on the opposite side. See Fig. 13-14. As in downhill welding, a root bead is deposited first after tack back and the weld is begun. While the root bead is being deposited, no electrode weaving motion is necessary. The electrode is simply advanced with a slow and uniform movement along the joint. As the electrode approaches the upper level, the molten metal begins to flow downward at a faster rate. When this happens, a slight whipping action is desirable to achieve better puddle control.

After the root bead is completed, one or more filler layers are deposited, followed by the final cover pass. Both fill and cover passes are made with a slant weave as described in downhill welding.

Electrodes

For most shielded metal-arc pipe welding, E-6010 or E-6011 electrodes are used. Where high tensile strength welds are required on criti-

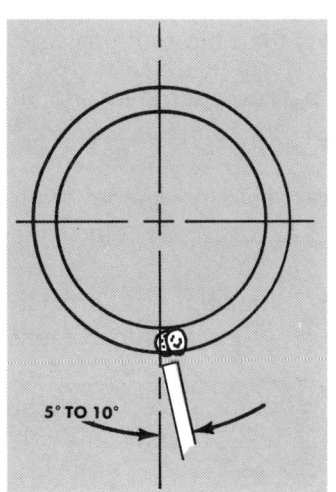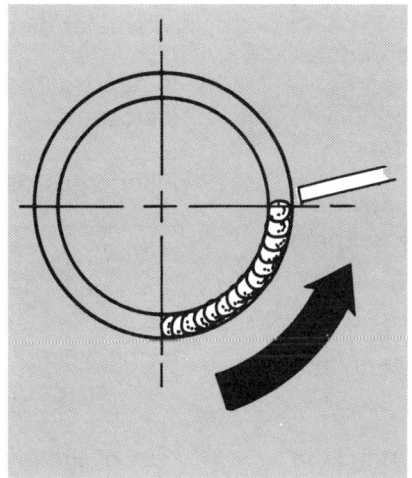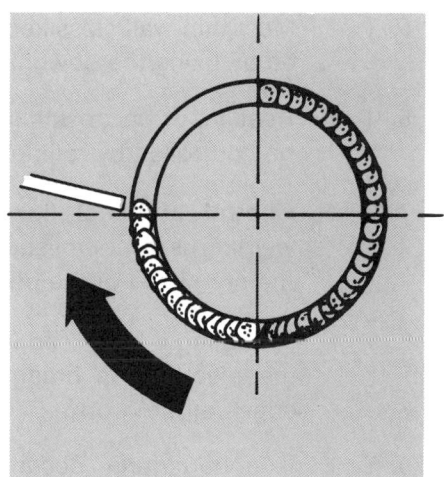

Fig. 13-14. Uphill welding.

welding. The weld is started just back of the bottom position or what is commonly known as the 6:30 position. Actually the arc is struck ahead of this spot and a long arc maintained for a short period to preheat the surface. Then it is brought cal pipe lines, electrodes in the E-70xx classification are used. As a rule, $5/32''$ diameter electrodes are recommended for the root bead and hot pass and $3/16''$ diameter electrodes for the filler and cover passes.

SELF QUIZ

Correct answers are listed in the back of the book.

True and False

Circle the letter T if the statement is true or F if it is false.

1. T F Small diameter pipes with wall thickness of 1/8" or less do not have to be beveled.

2. T F The included angle of beveled grooves for pipe run between 70 to 75 degrees.

3. T F Backing rings are used primarily to secure proper pipe alignment.

4. T F Tack welds on pipe should not penetrate to the root of the groove.

5. T F Most thin-wall classified pipe is welded by using the downhill technique.

6. T F Heat input on downhill welding can be controlled by rolling the pipe.

7. T F Control of metal flow in downhill welding is accomplished by keeping the arc ahead of the puddle.

8. T F In making a root bead in a pipe joint, the electrode is dragged along the edges of the groove.

9. T F If undercutting occurs during the process of depositing a root bead, increase the electrode travel speed.

10. T F The soundness of a pipe weld depends a great deal on the adequacy of the root bead.

11. T F In tying the ends of a pipe weld together, the electrode should be momentarily buried in the crater of the last bead, then raised slightly and the weld continued.

Fill-In

Supply the missing word or words where blanks appear.

12. To restart a pipe weld, the arc is struck about 1/2" _____ of the bead and then moved _____ with a _____ arc.

13. The specific function of the hot pass is to _____ out the remaining particles of _____ as well as to secure complete _____ of the base metal and root bead.

14. The hot pass is usually made with a _____ motion of the electrode.

15. The filler passes require _____ diameter electrodes than the root passes.

16. In depositing filler passes, each layer should start and end at a _____.

17. Filler passes are usually made with a _____ electrode motion.

18. The cap pass is intended to provide _____ reinforcement to the weld.

19. The cover pass should have a slight _____ above the surface of the pipe.

20. The _____ welding technique is used to weld heavy-wall pipe.

21. In uphill pipe welding the electrode is slanted about _____ degrees.

22. To start an uphill weld, the electrode is struck just past the _____ _____ position.

23. Uphill welding requires that a long arc be maintained momentarily ahead of the beginning bead to _____ the surface.

WELDING ASSIGNMENT

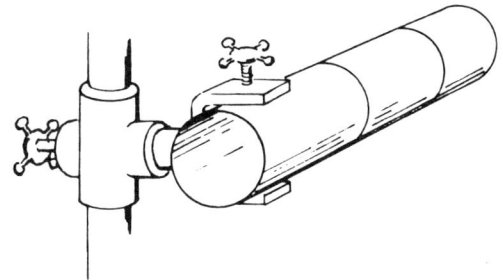

Fig. 13-15. Clamp practice pipe sections in a jig.

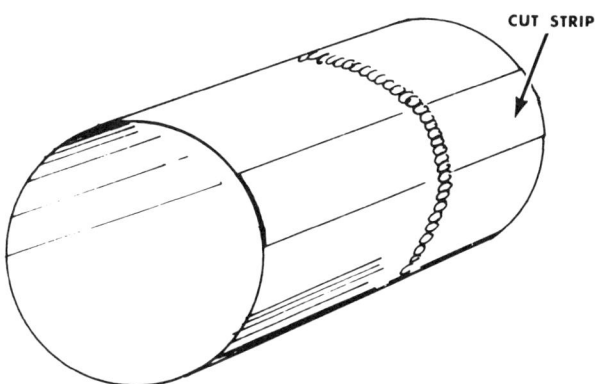

Fig. 13-16. Pipe test strip.

I. Welding Small Diameter Pipe with the Shielded Metal Arc—Single Pass.

1. Tack weld several lengths of pipe and clamp the assembly in a jig. See Fig. 13-15.
2. Make all welds with E-6010 or E-6011 electrodes.
3. Use a downhill welding technique as described in Unit 13.
4. To test the weld, cut a strip about 2″ wide from the pipe and then bend it in a vise. See Fig. 3-16.
5. Continue making single-pass welds on small diameter pipe until your instructor certifies that you have mastered this skill.

II. Welding Grooved Joints on Large Diameter Pipe with the Shielded Metal Arc

1. Secure several sections of large diameter pipe and bevel the edges. Ask your instructor what you should use to make the bevels.
2. Tack weld the pipe sections and then deposit a root bead. See Fig. 13-17.
3. Practice both downhill and uphill welding techniques in depositing filler passes.

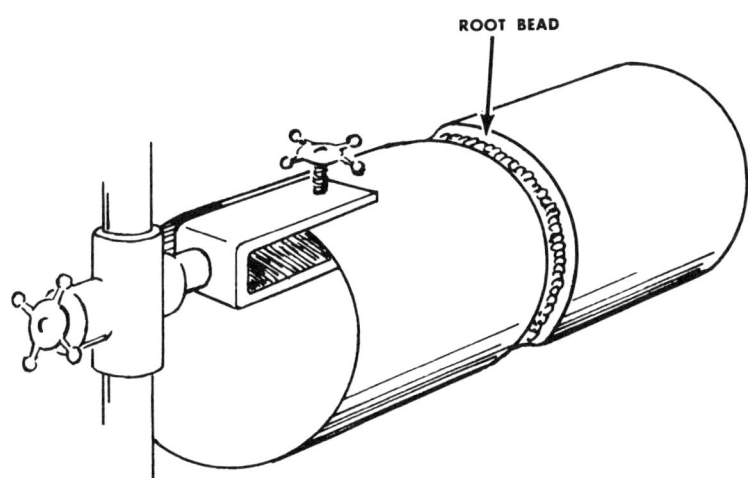

Fig. 13-17. Large diameter pipe practice welding.

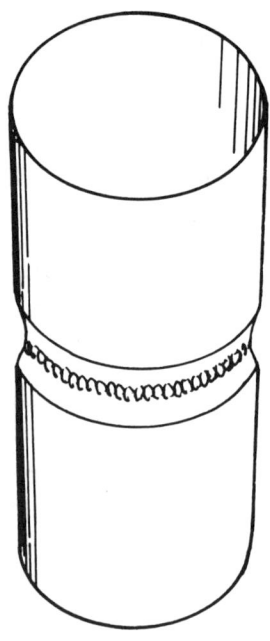

Fig. 13-18. Practice pipe weld in the horizontal position.

4. For some joints, place the pipe in a vertical position, so you can practice making horizontal welds around the pipe. See Fig. 13-18.

5. Cut strips from welded pipes and test them by bending them—as you did for the previous weld you made.

6. Continue to weld sections of large diameter pipe until you can produce sound welds.

III. Pipe Welding with Tig

1. Secure several lengths of small diameter pipe—low carbon steel, stainless steel, aluminum.

2. Tack them to form the joints shown in Fig. 13-19.

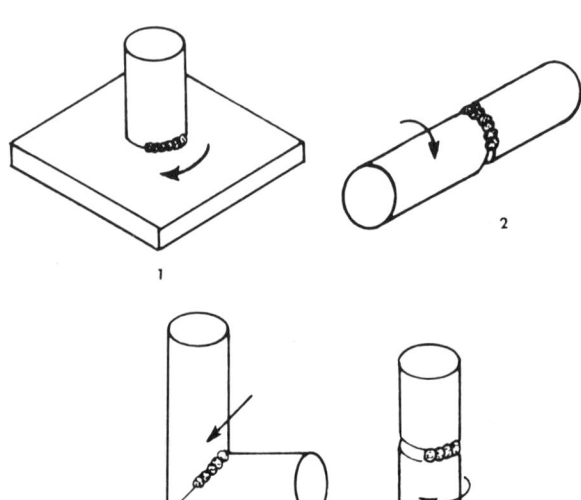

Fig. 13-19. Sequence of pipe welds with Tig.

3. Start to practice with low carbon steel pipe and then use stainless steel and aluminum.

4. Test each weld and evaluate the results. For this assignment, simply run a rough check on the welds by bending them in a vise.

IV. Pipe Welding with Mig

1. Select lengths of available large diameter pipe.

2. First practice depositing surface beads using both downhill and uphill techniques. See Fig. 13-20. Use E-70S-3 type wire in the 0.030"-0.035" diameter range.

3. Practice welding grooved pipe joints with the necessary number of passes. Follow the welding sequence shown in Fig. 13-20.

4. Test welds by cutting several specimens from the welded pipe (Fig. 13-21) and subject them to a bend test. Use the Weld Analysis Check List at the end of this unit to evaluate the test results.

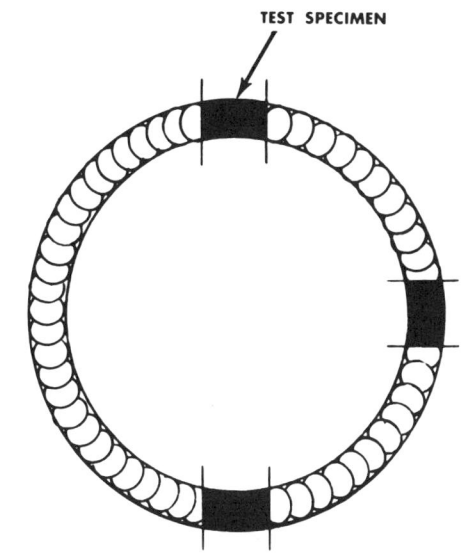

Fig. 13-21. Test specimens to be cut from welded pipe.

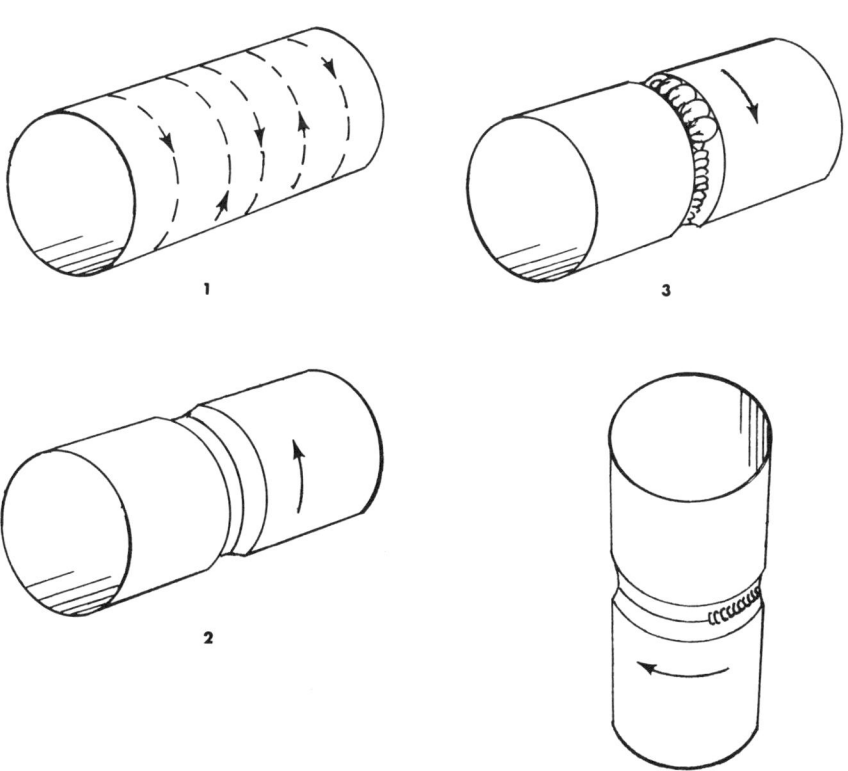

Fig. 13-20. Sequence of practice pipe welds with Mig.

Weld Analysis Check List

	Yes	No
1. Bead widths are right size	___	___
2. Beads have uniform ripples	___	___
3. Weld beads are too flat	___	___
4. Weld beads are too high	___	___
5. Weld penetration is insufficient	___	___
6. Weld penetration is excessive	___	___
7. Cold laps on surface	___	___
8. Weld has surface porosity	___	___
9. Weld has subsurface porosity	___	___
10. Weld has crater cracks	___	___
11. Weld has burn thru	___	___
12. End crater is filled	___	___
13. Weld passed bend test without cracking	___	___

Unit 14
Arc Cutting

Most metal cutting is generally done with the oxy-acetylene flame. However, there may be times when oxy-acetylene equipment is not available or the type of cutting is such that it can be more easily accomplished with the metallic arc. Whereas the flame cutting process depends on rapid oxidation of the metal, in shielded metal arc the cutting is simply a melting process produced by the extreme heat of the arc. The actual cutting operation is relatively simple, providing you follow a few basic procedures.

Cutting with the Shielded Metal Arc

The shielded metal arc does not produce as smooth and precise a cut as does the gas cutting torch, but cutting with the arc is fast and reasonably economical. The arc is particularly effective for cutting cast iron and steel for salvage purposes, small pieces, and areas which are hard to reach.

It is easy to understand why cutting with the arc is possible. The heat of the arc ranges from about 6500° to 10,000°F (about 3600° to 5500°C) whereas steel, for example, melts at about 2400°F (1315°C). Since cutting actually is a melting process, the excessive heat of the arc readily melts the metal.

Coated mild steel electrodes are used for cutting purposes with either AC or DC machines. Follow this procedure to carry out a cutting operation:

1. Set the machine for the same polarity as in welding for the electrode selected.

2. Use a mild steel coated electrode, such as E-6010 or E-6011. The diameter of the electrode will depend on the thickness of the metal to be cut and the amperage capacity of the machine. For most general purpose cutting, use $3/32''$ electrodes for metals up to $1/8''$ in thickness and $5/32''$ electrodes for materials which exceed $1/4''$ in thickness. Table 14-1 gives the approximate ampere setting for cutting.

TABLE 14-1. SUGGESTED AMPERE SETTING FOR CUTTING.

THICKNESS (inches)	ELECTRODE dia (inches)	CURRENT RANGE (amps)
Up to 1/8	3/32	75–100
Up to 1/8	1/8	125–140
Up to 1/4	5/32	140–180

Keep in mind that the amperage settings listed in Table 14-1 apply only to machines limited to 180 amperes. If you are using machines with higher capacity, you can employ larger diameter electrodes with higher ampere settings.

3. Place the metal in a flat position. Start the cut at the bottom edge of the plate, that is, the edge in the lowest position. When the diameter of the electrode is larger than the thickness of the plate being cut, simply move the electrode in a straight line. Check Fig. 14-1 and note the position of the electrode.

4. When the metal is heavier than the electrode, use a weaving motion to make the cut. Move the electrode with a quick upward motion and then push the electrode downward as shown in Fig. 14-2. The downward movement helps to force the molten metal out of the slot.

Fig. 14-3. Cut round stock from each side. (The Lincoln Electric Co.)

5. Flat stock over $1/8''$ in thickness is often easier to cut if placed in a vertical position.

6. To cut round stock, start the cutting at the outside edge so the metal can flow from the round bar. Carry the cut to the center of the piece and then start a new cut on the opposite side as shown in Fig. 14-3.

Piercing holes. To burn a small hole through thin stock, strike the arc and keep a long arc over the spot until the plate begins to sweat.

Next, bring the arc down into the molten pool of metal, moving the electrode in a circular motion as illustrated in Fig. 14-4. Continue the circular motion until the hole is pierced. It is a good idea to have a punch or bolt on hand so you can drive it into the hole while the metal is still hot. This will help true up the hole.

To pierce holes in metal over $1/4''$ in thickness, place the metal so the hole is made in a vertical position. This permits the molten metal to run

Fig. 14-1. To cut thin stock, move the electrode in a straight line. Use a guide to insure a straight cut.

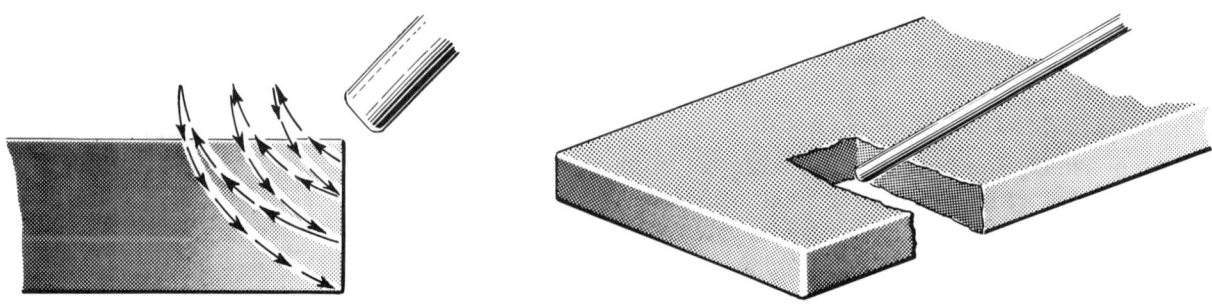

Fig. 14-2. Use a quick upward motion and then push the molten metal out with a downward motion.

Arc Cutting

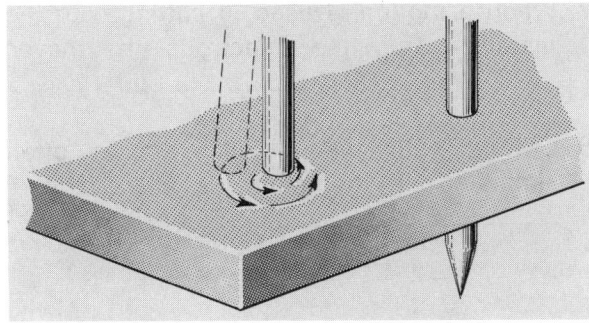

Fig. 14-4. To pierce a hole, bring the arc into the molten pool and move the electrode in a circular motion.

out of the hole as shown in Fig. 14-5. The metal that adheres to the outside surface is easily removed with a chisel or by the heat of the arc.

Cutting large holes. For cutting large holes, follow this procedure:

1. With a soapstone or center punch lay out the hole to be cut.

2. Pierce a small hole in the center of the area to be cut as shown in Fig. 14-6.

3. Move the electrode from the center around the edge of the marked circle.

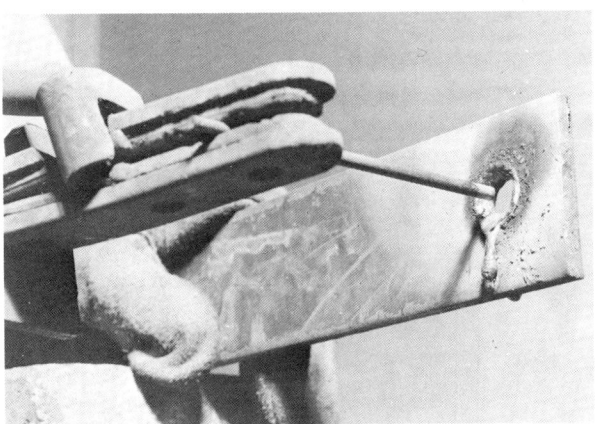

Fig. 14-5. To pierce holes in metal over 1/4", place the metal on edge. (The Lincoln Electric Co.)

Fig. 14-6. For large holes, start the cut in the center and move the electrode around the marked edges.

SELF QUIZ

Correct answers are listed in the back of the book.

Fill-In

Supply the missing word or words where blanks appear.

1. For cutting with the metallic arc the polarity of a DC machine depends on the type of _____ used.

2. The type of electrodes used for cutting are _____ or _____.

3. The diameter of the electrode used for cutting is governed by the _____ of the _____.

4. For cutting purposes the amperage must be set _____ than normally used for welding purposes.

5. When the electrode is larger than the thickness of the plate to be cut, the electrode should be moved without any _____ motion.

6. A weaving motion is used in cutting when the metal is _____ than the electrode.

7. When a motion is used in cutting with the metallic arc, the electrode is moved _____ and _____.

8. Cutting with the metallic arc is often simplified if the stock is placed in a _____ position.

CUTTING ASSIGNMENT

Practice cutting with the shielded metal arc. Use ³⁄₃₂″ diameter electrodes and ⅛″ mild steel plates. Make cuts as shown in Fig. 14-7.

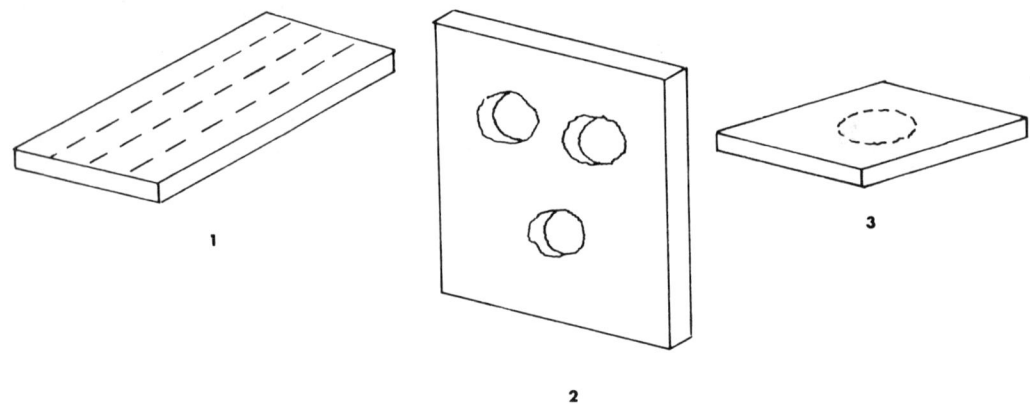

Fig. 14-7. Practice cutting sequence with the shielded metal arc.

UNIT 15
Certification of Welders

To protect human lives and to make certain that products or structures fabricated by welding will function safely and effectively, certain safeguards are generally established. The safeguards are usually stipulated in some document which clearly defines the nature and conditions of the required work.

The listing of specifications governing quality is not always as simple as it might appear because of the diversity of products and structures involving a great variety of welding processes and a wide gradation of welding skills. State and local laws sometimes specify in great detail what these requirements must be. At other times there may be only limited regulations, as established by the manufacturer, especially if high performance standards are not required. In all instances, safeguards relate directly to the performance skills of the welder. Degrees of welding competences may be expressed in the form of codes, standards or specifications.

Code

A code consists of a set of regulations covering permissible materials, service limitations, fabrication, inspection, testing procedures and qualifications of welding operators. These codes have been established by a number of nationally recognized agencies, such as

American Welding Society (AWS)
American Society of Mechanical Engineers (ASME)
American Petroleum Institute (API)
American National Standards Institute (ANSI)

Codes usually deal with a specific field of work such as ship building, piping, boiler work, building construction, tanks, aircraft and many others.

A code is sometimes enacted into law and consequently is often the most enforceable of any safety regulation.

Standards

Standards are specific regulations which cover the quality of a particular product to be fabricated by welding. By and large, standards deal with work quality rather than work procedure. Thus standards may cover type of material to be used, test strength of required welds, characteristics of filler metals, preheating and postheating temperatures and other essentials which have a direct bearing on the quality of the finished product.

Standards are usually developed by the manufacturer and apply only to its own welding personnel and work or product to be produced. The stringency of the standards depends on the nature of the work or product and the demands of the consumer. Thus for some jobs the required competency of the welders may not be unduly high, whereas for other tasks the performance requirements could be extremely critical. On many occasions the established standards are based on a nationally recognized code and may be supplemented by other demands which are dictated by the parties for whom the work is to be done.

Specifications

Specifications are specific descriptions of fabricating procedure. Among other manufacturing instructions, they include such welding data as location of welds, welding process to be used and method of testing the soundness of welds. These specifications are usually formulated by the design engineer and are included on production prints or on separate specification sheets. The nature of the specifications is governed by established standards and quality of work required. Welders must then produce the type of welds indicated by these specifications.

Certification Requirements

There is no one set of certification requirements dealing with all segments of the welding trade. Each area, whether it involves welding pipe, aircraft parts, building structures, boilers or ships, will have its own certification requirements.

Although certification requirements may vary somewhat for one classification of welders to another, in general, they all specify comparable tests which welders have to make. Most tests involve one or more of the following:

1. Tension test to establish the strength of a weld.
2. Guided bend tests to determine the ductility of the weld bead.
3. A fillet weld test to check the lack of fusion or cracks and proper weld contour.

These tests may have to be carried out in one or several positions, such as flat, horizontal, vertical, or overhead.

Qualifying tests are performed on specimens, sometimes called *coupons,* of the same material to be used in the product involved. For each weld test, pieces of certain size are selected and then subjected to some destructive or nondestructive forces. A welder is considered qualified only if his test specimens meet the required standards of quality.

Certifying Agency

Generally manufacturers have their own testing programs for qualifying welders. In such a program someone in the plant is designated as the certifying agent. A welder then reports to the agent and performs certain welding tasks that will demonstrate his or her ability to meet the requirements of the established standards.

The results of the test are analyzed and if found satisfactory are so stated and recorded by the company. The welder is then said to be qualified to work on the contracted job.

Certification Procedure

The question often arises among welders, "How can I become certified?" First of all one must remember that there are no standardized certification requirements which will lead to a general permanent certificate. Secondly, certification of welders is assumed by the manufacturer, or supplier who has contracted to provide certain types of products or perform special kinds of welding jobs. Consequently, as a welder you cannot go to just a single agency and apply for permanent certification. Each time you apply for a welding job, or if already employed each time you are placed on a different welding assignment, you have to subject yourself to some qualifying examination. If you pass the designated tests you are then certified for that particular job. If you move to some other assignment, you have to be certified again. For example, just because you are qualified to weld pipe does not automatically mean you are also qualified to weld boilers. Each time you have to be reexamined even though some of the actual tests may be similar to those you have previously taken.

Tensile Test. Tensile testing involves placing

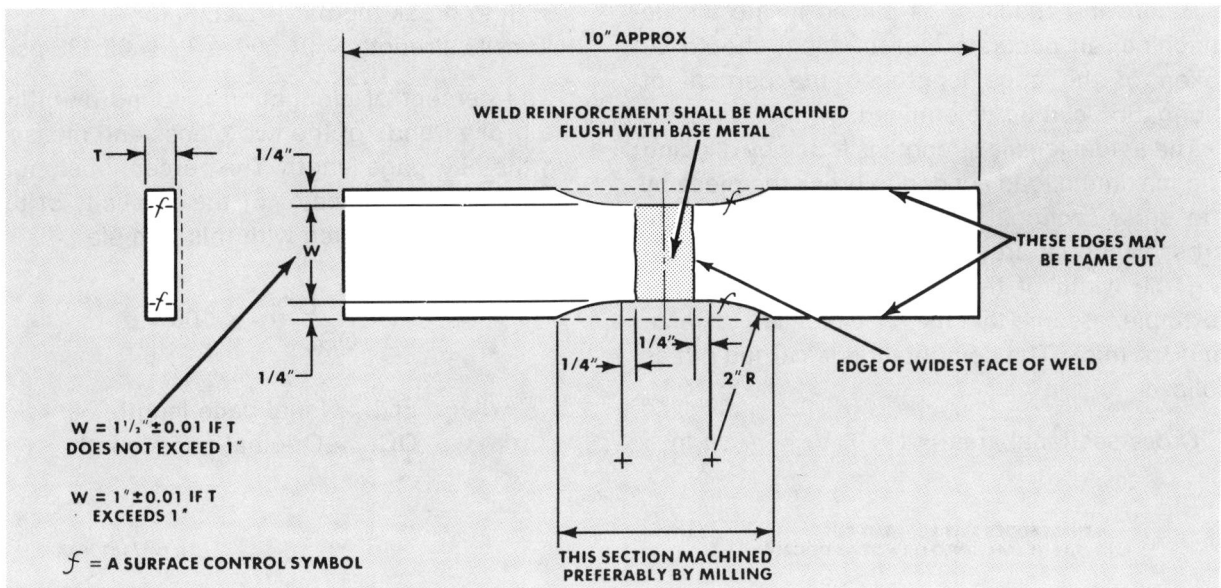

Fig. 15-1. Tensile specimen for flat plate butt weld. (AWS)

the weld specimen in a tensile testing machine and pulling the piece until it breaks.

The test specimen should be cut from a V butt joint having a 60° included angle. The weld plate should be of the same quality metal as specified for the product involved and welded with electrodes to produce the required tensile strength. See Figs. 15-1 and 15-2.

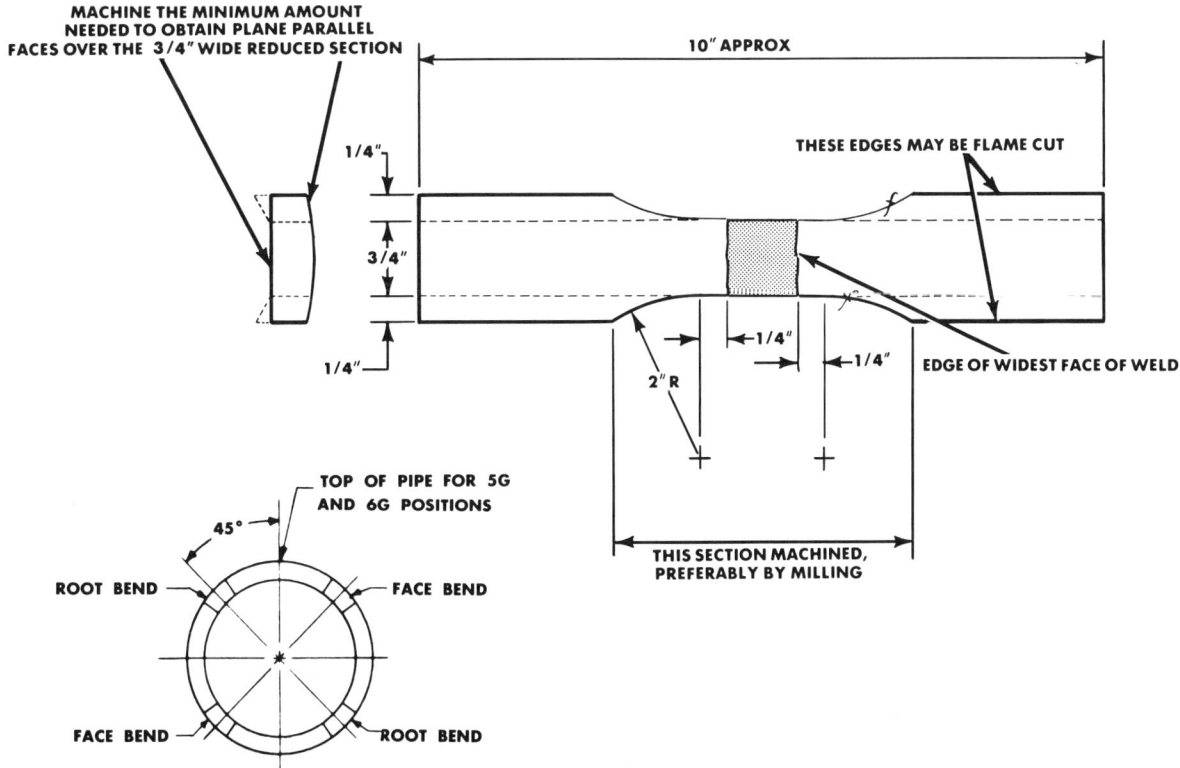

Fig. 15-2. Tensile specimen for pipe butt weld. (AWS)

Before the specimen is placed in the tensile machine, an accurate measurement should be taken of the gage length so the percent of elongation can be determined.

The actual tensile strength is found by dividing the maximum load needed to break the piece by the cross-sectional area of the specimen. The cross-sectional area is determined by multiplying the width of the bar by its thickness. For example, assume that the specimen is 1½" wide and ¼" thick. The computation is carried out as follows:

Cross-sectional area = 1½ × ¼ = ⅜ sq in

Pull to break the bar = 24,500 lb
Tensile strength = 24,500 ÷ ⅜ = 65,333 psi

The percent of elongation is found by fitting the broken ends of the two pieces and measuring the new gage length. The percent of elongation is a good indicator of the plasticity of the weld and is calculated with this formula:

$$\frac{FGL - OGL}{OGL} \times 100$$

where: FGL = Final gage length
OGL = Original gage length

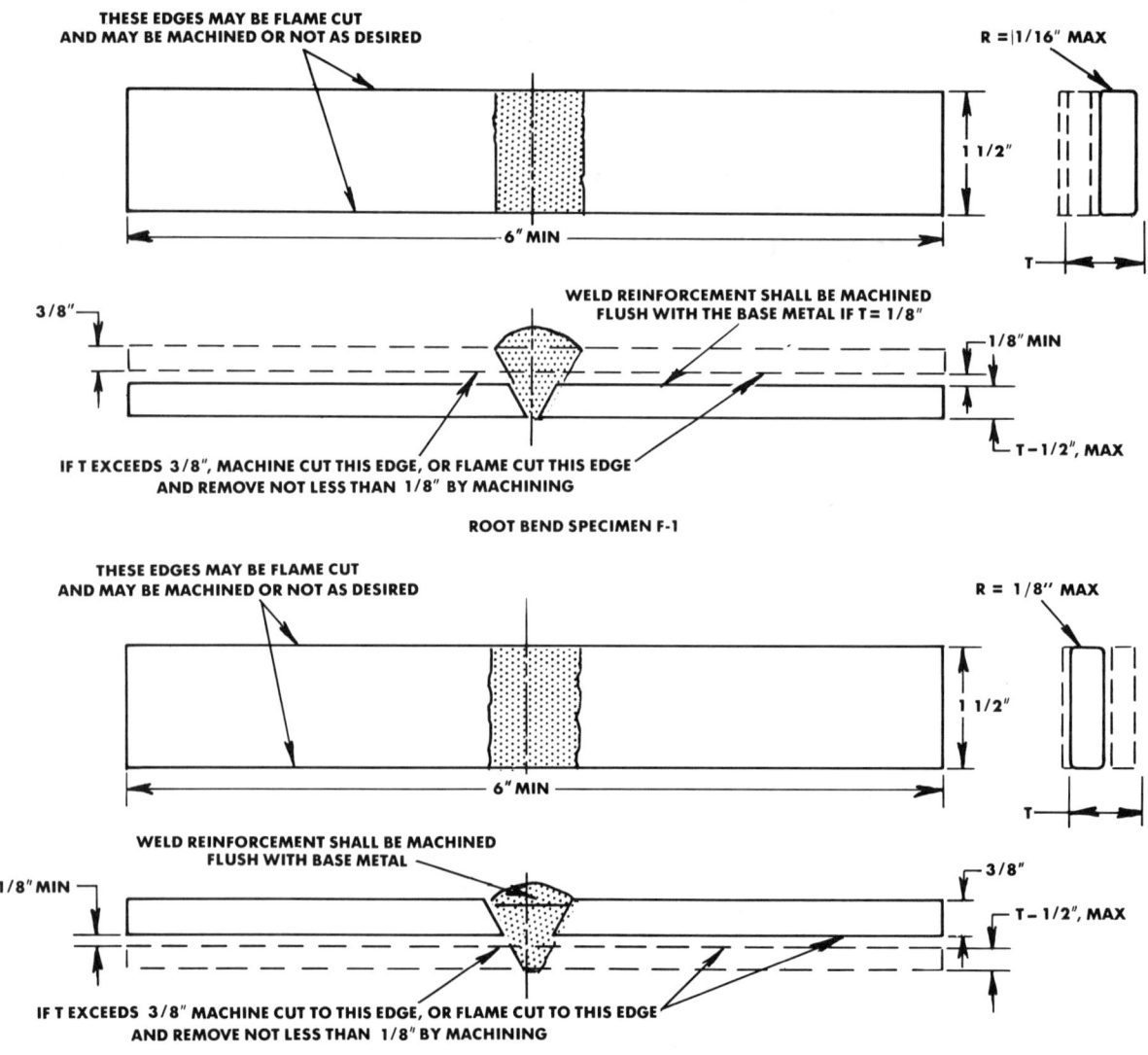

Fig. 15-3. Specimen for a guided bend test. (AWS)

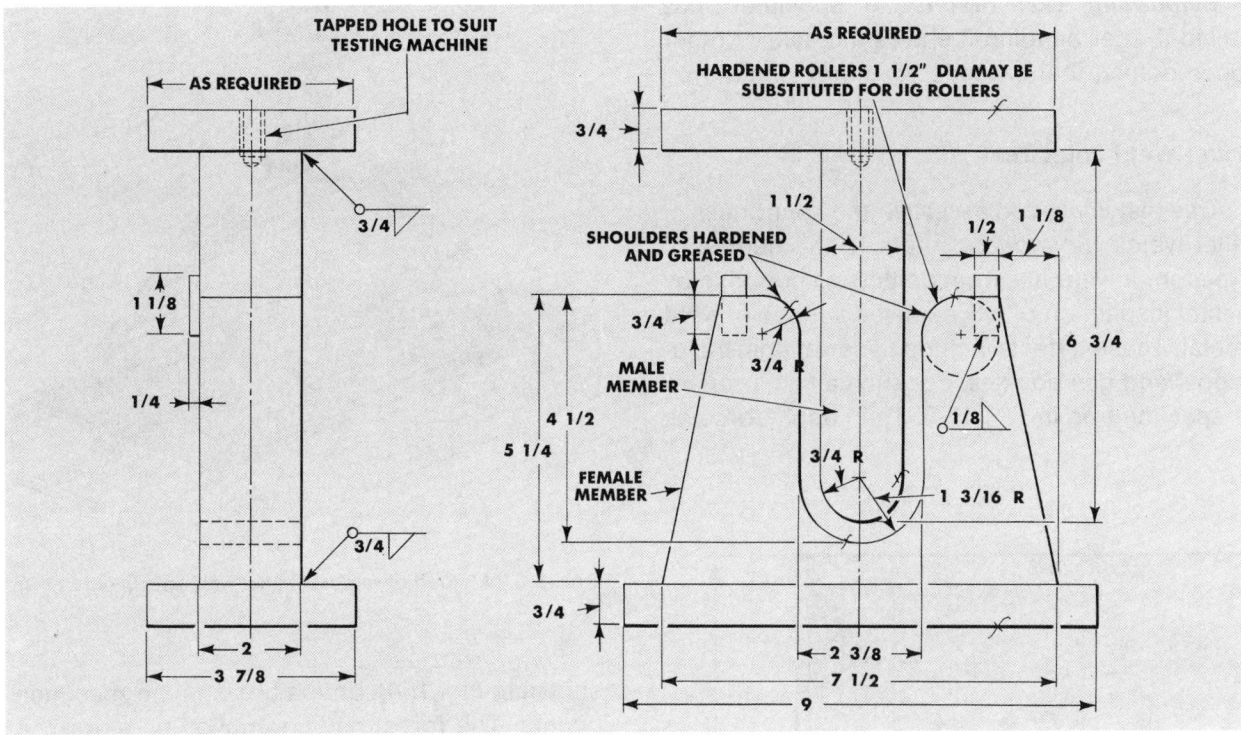

Fig. 15-4A. The detail working drawing of a jig for a guided bend test.

Acceptable test results. A test specimen should withstand a minimum load of 55,000 psi and a minimum elongation of 20% per 2″.

Guided bend test. For this test, two specimens, as shown in Fig. 15-3, are required. One piece referred to as the *face-bend specimen* is used to check the quality of fusion, that is, whether the weld is free of defects such as porosity, inclusions, etc. The second piece, referred to as the *root-bend specimen,* is used to check the degree of weld penetration.

To perform the face-bend test, place the specimen in the guided-bend jig face down and depress the plunger until the piece becomes U-shaped in the die. See Fig. 15-4A and 15-4B.

In the root-bend test, place the specimen in the jig with the root down or in just the reverse position of the face-bend piece. To be acceptable, the results must also show no cracks.

Fig. 15-4B. A jig being used to perform a test.

Evaluating test results. A specimen has failed if after bending it shows a crack, or other open defect, that exceeds 1/8" in any direction.

Fillet Weld Joint Test

This test is used to ascertain the soundness of fillet welds. *Soundness* refers to the degree of freedom a weld has from defects discernable by visual inspection of any exposed surface of weld metal. These defects include penetrations, inclusions, and gas pockets. For such a test, prepare a specimen as in Fig. 15-5. Then apply force as

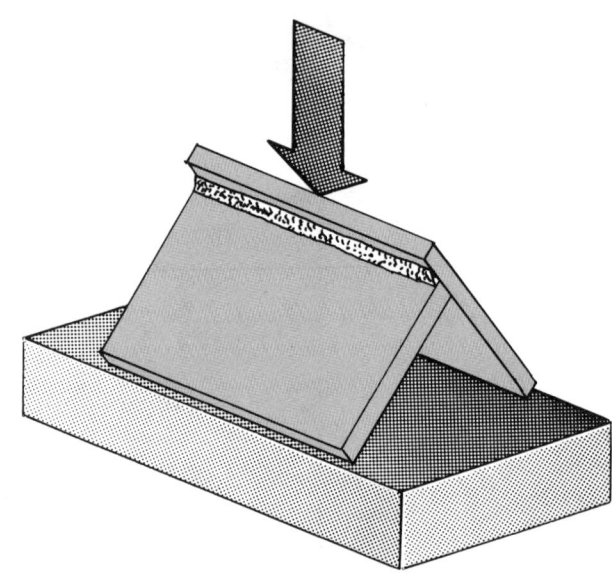

Fig. 15-6. Method of rupturing fillet weld specimen. (AWS)

shown in Fig. 15-6, until a break in the specimen occurs. The force may be applied by a press, a testing machine, or hammer blows.

Evaluating test results. To pass visual inspection, a test specimen must meet the following requirements:
1. Weld must be free of cracks.
2. Crater must be filled and free of cracks.
3. Fillet must have an acceptable profile, as shown in Fig. 15-7.

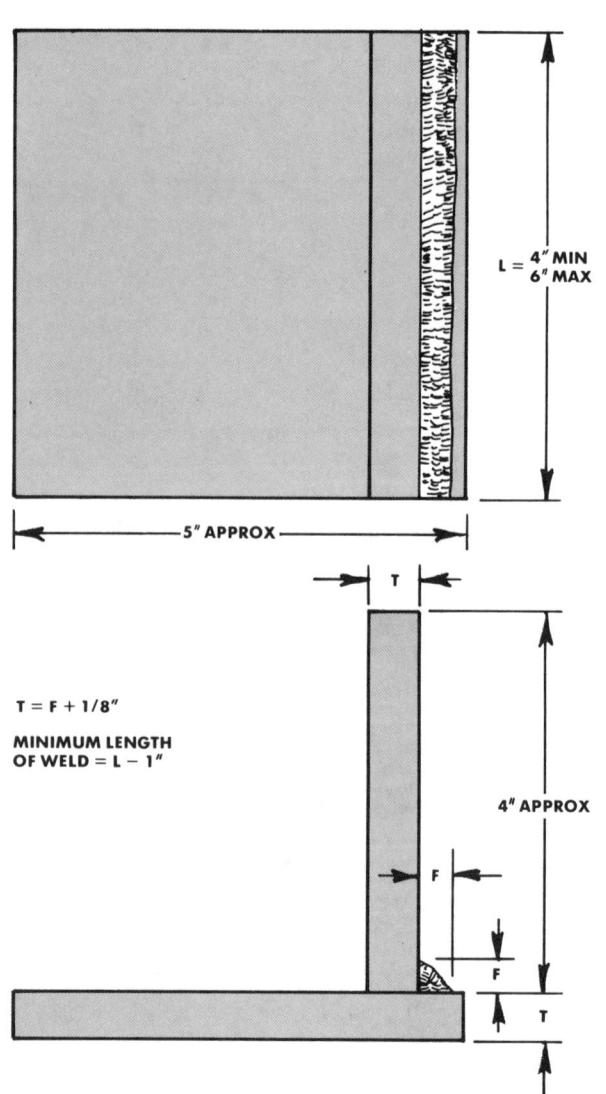

Fig. 15-5. Specimen for a fillet weld test. (AWS)

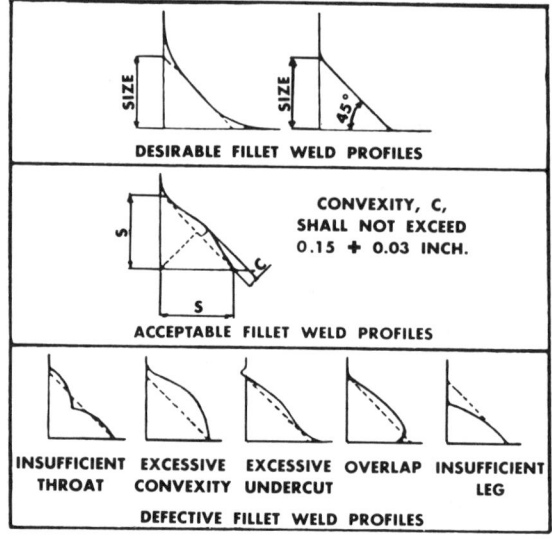

Fig. 15-7. Weld profiles. (AWS)

STUDENT CERTIFICATION TEST

Upon completion of this welding course, each student should be required to take one or all of these qualifying tests: tensile, guided bend, and fillet. For students to conduct these tests, the instructor should provide the following specifications:

1. Type of test
2. Type of metal
3. Thickness of metal
4. Welding method (SMAW, GTAW, GMAW)
5. Type and diameter of electrode
6. Welding position
7. Weld bead size
8. Number of specimens

Answers to Self Quiz Items

Unit 1

1. T
2. F
3. F
4. T
5. F
6. F
7. bead
8. crater
9. pass
10. ripples
11. fillet
12. A. penetration
    B. bead
    C. base metal
    D. reinforcement
    E. root face
    F. root opening
    G. leg
    H. toe
    I. face
    J. throat
    K. toe
    L. leg
    M. root
    N. surface weld
    O. fillet
    P. square butt joint
    R. V-groove joint

Unit 2

1. A. flange-edge
    B. flange corner
    C. square groove
    D. V groove
    E. bevel groove
    F. U groove
    G. J groove
    H. flare V-groove
    I. flare bevel-groove
    J. fillet
2. A. back or backing
    B. melt thru
    C. surfacing
3. A. weld all around
    B. field weld
    C. flush contour
    D. convex contour
    E. concave contour
4. A. arrow side
    B. other side
    C. both sides
    D. both sides, 2 joints
5. $3/8''$ fillet weld, both sides 10'' long
6. A. $1/4''$ fillet welds on both sides
    B. arrow side fillet weld $5/16''$, other side fillet weld $1/4''$
7. $3/8''$ fillet weld, arrow side, 6'' long, 10'' pitch
8. combined weld – bevel groove weld followed by fillet weld on arrow side; bevel groove weld on other side; horizontal member to be beveled
9. fillet weld has unequal legs
10. A. square groove weld, $1/16''$ root opening
    B. V-groove weld with 50° bevel angle
11. flange-edge joint, $1/16''$ radius, $1/8''$ height of flange, $3/32''$ weld size
12. weld surface to be built up $3/32''$ height
13. V-groove joint, arrow side, weld to be convex by grinding
14. square groove, arrow side, complete melt thru, $1/16''$ reinforcement
15. staggered intermittent weld, 3'' long and 6'' pitch
16. member to be beveled

Unit 3

1. c
2. b

3. c
4. b
5. c
6. c
7. a
8. d
9. b
10. c
11. stress
12. strain
13. elastic limit
14. tensile strength
15. cast iron
16. torsional strength
17. toughness
18. malleable

Unit 4

1. T
2. T
3. T
4. T
5. F
6. T
7. F
8. F
9. F
10. T
11. F
12. T
13. T
14. T
15. F
16. T
17. F
18. direct current, reverse polarity
19. direct current, straight polarity
    alternating current
    flat position
    vertical position
    overhead position
    horizontal position
19. electrode
    tensile strength
    welding position
    manufacturer's characteristics
20. A. Fast-fill-----E-6020—F, H positions —deep penetration
    E-7024—F, H positions—high deposition rate
    B. Fast-freeze-----E-6010, E-6011—all positions—all purpose
    C. Fill-freeze---E6012, E6013---general purpose repair work, vertical and overhead, poor fit-up
    D. Low hydrogen---E7018, E7024---high sulfur, high carbon steels

Unit 5

1. DC
2. amperage
3. volt
4. open circuit voltage
5. 18-36
6. polarity
7. reverse
8. straight
9. straight
10. AC
11. true
12. true
13. ultraviolet and infrared
14. amperage
15. goggles
16. A flash will occur.

Unit 6

1. cracking
2. globules
3. penetration
4. sticks, high
5. diameter
6. atmosphere
7. melts
8. heat
9. narrow
10. wide, high
11. penetration
12. arc
13. lift, side, ahead
14. filled
15. high
16. current, low

Unit 7

1. A. butt
   B. lap
   C. edge
   D. T
   E. corner
2. A. closed
   B. open
   C. V
3. c
4. b
5. d
6. a
7. c
8. b
9. d
10. c
11. d
12. c
13. b
14. T
15. T
16. F
17. F
18. F
19. F
20. F

Unit 8

1. T
2. T
3. F
4. F
5. T
6. T
7. F
8. T
9. T
10. F

Unit 9

1. formation of overlaps
2. Penetration is shallower, danger of burn thru is less, and deposition rate is faster.
3. 10°–15°
4. no
5. by a slight crescent motion
6. because of better penetration
7. by using a rocking or whipping motion
8. to reduce heat and prevent puddle from sagging
9. twice the diameter of the electrode
10. uphill—fast-freeze E6010, E7014
    downhill—fill-freeze E6012, E7014

Unit 10

1. fast freeze
2. downward
3. yes
4. over the shoulder
5. rocking the electrode
6. $3/16''$
7. 10°–15°
8. to keep heat low or puddle small
9. on the side of the arc
10. to protect the welder from sparks

Unit 11

1. horizontal, $1/8''$
2. touch
3. horizontal
4. 75°
5. circular
6. leading edge
7. argon
8. arc
9. gas tungsten arc welding
10. alternating current, high frequency
11. direct current, reverse polarity
12. larger
13. DCSP
14. reverse
15. AC
16. A. cubic feet per hour
    B. liters per minute
17. F
18. T
19. F
20. F
21. T
22. F
23. T
24. T
25. F
26. F
27. T
28. F
29. F
30. F
31. F
32. T
33. T
34. T
35. F
36. T
37. T
38. F

Unit 12

1. a
2. c
3. b
4. d
5. a
6. c
7. a
8. b
9. a
10. d
11. b
12. c
13. T
14. F
15. T
16. F
17. T
18. F
19. T
20. T
21. T
22. T
23. pulled
24. forward
25. wire speed
26. run-in, scratch
27. one inch
28. reverse, speed

Unit 13

1. T
2. T
3. F
4. F
5. T
6. F

7. T
8. T
9. F
10. T
11. T
12. back, forward, long
13. burn, slag, fusion
14. whipping
15. larger
16. different point
17. slanting
18. maximum
19. crown
20. uphill
21. 5°–10°
22. 6:30
23. preheat

Unit 14

1. electrode
2. E-6010, E-6011
3. thickness, metal
4. higher
5. weaving
6. thicker
7. upward, downward
8. vertical